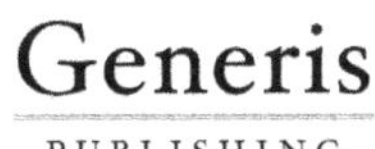

AF505508

The Vitalities of Ecological and Socio-Economic Integrities of Forests in Cities

Adili Yohana Zella
Luzabeth Jackson Kitali

Title: *The Vitalities of Ecological and Socio-Economic Integrities of Forests in Cities*

ISBN: 979-8-89248-812-9

Author: Adili Yohana Zella, Luzabeth Jackson Kitali

Cover image: https://pixabay.com/

Publisher: Generis Publishing
Online orders: www.generis-publishing.com
Contact email: info@generis-publishing.com

AUTHORS

[1]*Adili Y. Zella and* [2]*Luzabeth J. Kitali,*

[1]*Department of Economics, Faculty of Leadership and Management Sciences,*
The Mwalimu Nyerere Memorial Academy, P.O Box 9193, Dar es Salaam, Tanzania.
Tel: +255 22 2820041; +255 787 260448;
E-mail: zellahadil@gmail.com; adil.zellah@mnma.ac.tz

[2]*Department of Geography and History, Faculty of Arts and Social Sciences,*
The Mwalimu Nyerere Memorial Academy, P.O Box 9193, Dar es Salaam, Tanzania.
Tel: +255 22 2820041; +255 787 403734;
E-mail: bettyjm77@gmail.com; luzabeth.kitali@mnma.ac.tz

TABLE OF CONTENTS

PART I: INTRODUCTION AND BACKGROUND

CHAPTER ONE

1. INTRODUCTION TO URBAN FORESTS

1.1 Definition, Scope, and Importance of Forests in Cities

Forests are ecosystems defined by dense tree coverage and high biodiversity, offering essential ecological, economic, and social functions. Forests and urban green spaces in cities are essential for maintaining environmental balance, reducing climate change impacts, and improving urban livability. Urban forests include natural forests, planted woodlands, street trees, and green corridors located within or adjacent to urban areas, constituting an essential element of green infrastructure (Food and Agriculture Organization [FAO], 2023). Their significance extends beyond ecological limits, playing a crucial role in urban resilience, socio-economic advancement, and public health.

1.1.1 Global Perspective

Urban forests are acknowledged worldwide for their diverse functions. They regulate climate by sequestering carbon dioxide and moderating temperatures, functioning as natural air conditioners in urban heat islands (United Nations Environment Programme [UNEP], 2022). Urban forests improve air quality through pollutant filtration, stabilize soils, decrease noise pollution, and manage stormwater, thereby mitigating the effects of urban flooding (Nowak et al., 2023).

Urban forests enhance quality of life through the provision of recreational spaces, support for mental health, and an increase in property values from a socio-economic standpoint. They promote environmental education and community engagement, enhancing the inclusivity and sustainability of urban areas (Konijnendijk et al., 2022). Urban forests, while beneficial, face mounting pressures from rapid urbanization, pollution, and climate change, highlighting the need for integrated approaches to forest conservation within city planning.

1.1.2 Africa: Urban Forests Amid Rapid Urbanization

Forests in Africa are essential for the sustenance of urban and peri-urban ecosystems. Urban forests play a crucial role in addressing the challenges associated with rapid urbanization, including air pollution, heat stress, and water scarcity. Cities such as Nairobi, Kenya, have incorporated green spaces like Karura Forest into urban planning, illustrating the role of forests as ecological and recreational centers (Wanjiru & Matsui, 2022).

Africa's urban forests enhance socio-economic resilience through the provision of fuelwood, medicinal plants, and employment opportunities. Nonetheless, encroachment, deforestation, and insufficient policies for the sustainable management of urban green spaces pose significant threats to them (FAO, 2023). Enhancing governance and community engagement is crucial for the preservation of urban forests and their associated benefits throughout the continent.

1.1.3 Sub-Saharan Africa: Forests and Urban Challenges

Urban forests in Sub-Saharan Africa function as essential resources for cities experiencing rapid growth. They offer essential ecosystem services such as air purification, carbon sequestration, and biodiversity conservation. Addis Ababa, Ethiopia, has implemented urban reforestation initiatives to address deforestation and improve urban resilience (African Development Bank [AfDB], 2022).

The growth of informal settlements and dependence on forest resources for energy present considerable challenges. Urban planning frameworks frequently overlook the significance of forests, resulting in their deterioration. Integrating forests into urban development plans and fostering public-private partnerships is essential for sustainable management (UNEP, 2022).

1.1.4 East Africa: Forests in the Face of Urbanization

East Africa hosts distinctive forest ecosystems, such as the Eastern Arc Mountains and coastal forests, recognized as biodiversity hotspots and vital natural resources for urban areas (Myers et al., 2023). Urban forests in cities such as Kampala, Uganda, and Dar es Salaam, Tanzania, play a vital role in reducing urban heat, managing stormwater, and offering recreational green spaces.

Urbanization and population growth have exerted significant pressure on these ecosystems. The Green Belt Movement in Kenya exemplifies the effectiveness of community-driven conservation initiatives in the restoration and protection of urban forests (Maathai, 2021). Expanding these initiatives throughout East Africa can facilitate the incorporation of forests into urban planning and improve ecological and socio-economic resilience.

1.1.5 Tanzania: Urban Forests and Sustainable Development

The urban forests of Tanzania, particularly in Dar es Salaam and Dodoma, play a crucial role in maintaining ecological stability and supporting socio-economic development. Urban and peri-urban forests offer essential ecosystem services, including temperature regulation, air purification, and support for biodiversity (Ministry of Natural Resources and Tourism [MNRT], 2022). They contribute to the livelihoods of urban and peri-urban communities by providing firewood, construction materials, and traditional medicine.

Urban forests such as Pugu and Kazimzumbwi Forest Reserves adjacent to Dar es Salaam function as essential green buffers, mitigating environmental degradation and providing areas for recreation and education (Nyika, 2020). These forests encounter challenges including illegal logging, agricultural encroachment, and insufficient funding for conservation efforts.

1.1.6 Kazimzumbwi Forest Reserve: A Case Study of Urban Forest Integrity

Kazimzumbwi Forest Reserve, situated in the peri-urban region of Dar es Salaam, illustrates the significance of urban forests for ecological and socio-economic stability. Founded in 1936, it covers an area of 4,887 hectares and ranks among Tanzania's oldest coastal forests. The closeness to Tanzania's largest city renders it an essential natural resource for urban resilience and sustainability (Mwakalila et al., 2023).

Ecological Importance

Kazimzumbwi Forest Reserve hosts a variety of flora and fauna, encompassing numerous endemic and threatened species. It is essential for carbon sequestration, groundwater recharge, and temperature regulation, thereby alleviating the impacts of urban heat islands (Mgaya & Samwel, 2021). The preservation of this entity is essential for sustaining ecological stability in Dar es Salaam.

Socio-economic Contributions

The forest supplies vital resources to adjacent communities, such as firewood, honey, and medicinal plants. The initiative provides avenues for ecotourism, environmental education, and recreation, thereby supporting the local economy and improving public health (Nyika, 2020). Incorporating Kazimzumbwi into urban planning can enhance its socio-economic advantages while promoting sustainable urban development.

Challenges and Conservation Efforts

Kazimzumbwi encounters considerable challenges, such as deforestation, encroachment, and inadequate enforcement of conservation policies. Community-based conservation programs and government initiatives, including the National Forest Policy of 1998, seek to address these issues (MNRT, 2022). Enhancing these efforts can improve the forest's contribution to the ecological and socio-economic integrity of Dar es Salaam.

1.2 Historical Context of Urban Forests in City Planning

Urban forests have played a crucial role in city planning for centuries, transitioning from mere aesthetic features of ancient cities to essential elements of urban resilience and sustainability. The historical context of urban forests illustrates the evolving priorities of human settlements, transitioning from beautification and recreation to tackling ecological and socio-economic challenges. This section analyzes the historical evolution of urban forests worldwide, with a specific emphasis on Africa, Sub-Saharan Africa, East Africa, Tanzania, and the Kazimzumbwi Forest Reserve.

1.2.1 Global Perspective

Urban forests were intentionally integrated into city planning during the Renaissance period. Cities such as Florence and Paris integrated tree-lined boulevards and parks into their designs, embodying aesthetic principles and a pursuit of harmony with nature (Konijnendijk, 2008). During the 19th century, industrialization catalyzed swift urbanization, resulting in densely populated and polluted urban areas. Urban forests and parks address these challenges by enhancing public health and facilitating recreation. New York's Central Park, designed in the 1850s, was intended as a green refuge for city residents (Lawrence, 2020).

During the mid-20th century, there was an increasing acknowledgment of the ecological roles of urban forests, including air purification, temperature regulation, and flood mitigation (FAO, 2023). The emergence of global environmental movements in the 1970s, supported by the United Nations Habitat I Conference in 1976, highlighted the significance of urban forests in sustainable urban planning. Urban forests play a crucial role in global initiatives such as the Sustainable Development Goals (SDGs), especially SDG 11, which focuses on Sustainable Cities and Communities.

1.2.2 Africa: The Role of Urban Forests Amid Rapid Urbanization

The historical incorporation of forests into urban planning in Africa is closely linked to colonial and post-colonial land management practices. In the colonial era, urban green spaces primarily catered to the recreational needs of European settlers, while access for local populations was restricted. Examples include botanical gardens established in urban areas such as Nairobi and Cape Town (Wanjiru & Matsui, 2022).

Following independence, African nations prioritized the socio-economic and ecological advantages of urban forests. Rapid urbanization frequently surpasses planning initiatives, resulting in the deterioration of urban forests (African Development Bank [AfDB], 2022). Currently, initiatives such as the African Forest Landscape Restoration Initiative (AFR100) focus on restoring degraded landscapes, including urban forests, to promote sustainable urban development (UNEP, 2022).

1.2.3 Sub-Saharan Africa: Balancing Development and Conservation

Urban forests in Sub-Saharan Africa have been historically overlooked in urban planning processes. Their significance has increased due to urban challenges, including air pollution, heat islands, and flooding. Urban forestry has been incorporated into the planning frameworks of cities such as Addis Ababa in Ethiopia and Kampala in Uganda to tackle these challenges (Myers et al., 2023).

In the 20th century, Sub-Saharan Africa underwent significant deforestation attributed to urban expansion and dependence on wood as an energy source. Urban forests, frequently perceived as impediments to development, were removed to facilitate infrastructure (FAO, 2023). In recent decades, the significance of urban forests has undergone re-evaluation, with community-based initiatives and policy frameworks aiming to incorporate them into urban planning.

1.2.4 East Africa: Forests in the Context of Urban Development

Urban forests in East Africa, particularly in cities such as Nairobi and Dar es Salaam, have traditionally functioned as ecological buffers and livelihood sources. Colonial administrations created urban green spaces, exemplified by Nairobi's Karura Forest, to preserve biodiversity and manage urban climates (Maathai, 2021).

Following independence, urban forests encountered considerable challenges, such as encroachment and deforestation. Countries such as Kenya have initiated extensive reforestation projects, highlighting the socio-economic and ecological advantages of urban forests (Wanjiru & Matsui, 2022). These initiatives underscore the capacity of urban forests to facilitate sustainable urban development and alleviate the effects of climate change.

1.2.5 Tanzania: Historical Integration of Forests in City Planning

The development of urban forests in Tanzania has been influenced by its colonial legacy and subsequent post-independence policies. In the German and British colonial periods, forest management focused mainly on timber production and agricultural expansion, neglecting their ecological and urban functions (Kajembe & Malimbwi, 2021). Urban green spaces were frequently created for colonial administrators, resulting in restricted access for local communities.

Following independence in 1961, Tanzania prioritized the equilibrium between conservation and development. The National Forest Policy of 1953, updated in 1998, highlighted the importance of sustainable forest management and the involvement of communities (Ministry of Natural Resources and Tourism [MNRT], 2022). Urban forests have been consistently under-prioritized in city planning, resulting in the degradation of peri-urban forests such as Pugu and Kazimzumbwi.

The Tanzanian government has acknowledged the significance of urban forests in mitigating urban issues, including flooding, heat stress, and air pollution. The Tanzania Urban Resilience Program (TURP) seeks to incorporate green infrastructure, such as urban forests, into urban planning frameworks (Nyika, 2020).

1.2.6 Kazimzumbwi Forest Reserve: A Case Study

The Kazimzumbwi Forest Reserve, situated on the periphery of Dar es Salaam, illustrates the historical and modern difficulties associated with incorporating forests into urban planning. Kazimzumbwi was established in 1936 during the colonial era and was initially designated as a timber resource (Mwakalila et al., 2023). Over time, it evolved into a vital ecological resource, offering habitat for various flora and fauna while functioning as a carbon sink.

Historical Role

In the mid-20th century, Kazimzumbwi experienced significant logging and encroachment, indicative of wider patterns in forest management amid Tanzania's industrialization and urbanization (Mgaya & Samwel, 2021). Although classified as a forest reserve, insufficient enforcement of conservation policies permitted the continuation of illegal activities.

Contemporary Challenges and Opportunities

Currently, Kazimzumbwi encounters issues including deforestation, urban encroachment, and insufficient funding for conservation efforts. It also offers opportunities for the integration of urban forestry into the city planning of Dar es Salaam. Community-based conservation initiatives, with support from government and international organizations, seek to restore degraded areas and improve the ecological and socio-economic functions of forests (MNRT, 2022).

Integration into Urban Planning

Kazimzumbwi's closeness to Dar es Salaam establishes it as an essential green buffer that mitigates urban heat, flooding, and air pollution. The incorporation of this element in urban planning frameworks can improve resilience, offer recreational areas, and aid in the conservation of biodiversity. The integration of Kazimzumbwi into the urban development plans of Dar es Salaam underscores the increasing acknowledgment of forests as essential elements of sustainable urban environments (Nyika, 2020).

1.3 Overview of Ecological and Socio-economic Integrities

Forests, encompassing urban and peri-urban regions, are essential ecosystems that offer significant ecological and socio-economic advantages. They maintain biodiversity, regulate climate, and support livelihoods, constituting a crucial element of sustainable development (Food and Agriculture Organization [FAO], 2023). This section examines the integrity of forests, starting with a global overview and focusing on the Kazimzumbwi Forest Reserve in Tanzania.

1.3.1 Global Perspective

Forests globally encompass around 31% of the Earth's land area, delivering critical ecosystem services vital for sustaining life on the planet (FAO, 2022). They function as carbon sinks, sequestering approximately 7.6 billion metric tons of CO_2 each year, thereby contributing to climate change mitigation (United Nations Environment Programme [UNEP], 2022). Forests play a crucial role in maintaining hydrological cycles, regulating both local and global climates, and preventing soil erosion, thereby contributing to ecological stability (IPCC, 2021).

Forests play a crucial role in the global economy from a socio-economic perspective. They employ over 13 million individuals globally and provide resources including timber, non-timber products, and medicinal plants (FAO, 2023). Forest-based ecotourism generates significant annual revenue, highlighting the dual function of forests in conservation and economic development.

Notwithstanding these advantages, deforestation and forest degradation represent considerable threats. Since 1990, urbanization, agriculture, and infrastructure development have resulted in the loss of roughly 420 million hectares of forest (FAO, 2022). This loss highlights the necessity of incorporating forests into urban and peri-urban planning.

1.3.2 Africa: A Continent of Biodiversity and Dependence

Africa contains some of the most diverse and significant forest ecosystems, notably the Congo Basin, which is the second-largest tropical rainforest in the world (African Forest Forum, 2022). African forests serve a vital ecological function, sequestering approximately 66 gigatons of carbon and supporting numerous endemic species (World Resources Institute [WRI], 2022).

Forests in Africa play a crucial role in the socio-economic sustenance of millions. More than 70% of Africans depend on forests for energy, food, and medicine (FAO, 2023). Forests in urban areas deliver critical ecosystem services, including mitigation of the urban heat island effect and enhancement of air quality. Deforestation rates in Africa are among the highest in the world, primarily due to agricultural expansion, logging, and urban sprawl (UNEP, 2022).

1.3.3 Sub-Saharan Africa: Balancing Development and Conservation

Forests in Sub-Saharan Africa are essential to rural and urban communities alike. They play a crucial role in agriculture, water management, and biodiversity conservation (African Development Bank [AfDB], 2022). Urban forests, frequently overlooked in planning processes, are increasingly acknowledged for their contribution to urban resilience in the face of climate change.

Urban forest projects in Sub-Saharan Africa, including reforestation initiatives in Nairobi, Kenya, illustrate the increasing recognition of the ecological and socio-economic importance of forests (Wanjiru & Matsui, 2022). Limited funding and inadequate governance impede effective forest management in numerous urban areas.

1.3.4 East Africa: Forests and Urbanization

The forests of East Africa, encompassing the Eastern Arc Mountains and coastal regions, represent significant biodiversity hotspots that deliver essential ecosystem services (Myers et al., 2023). Urban forests in Dar es Salaam, Tanzania, play a crucial role in flood mitigation, air quality improvement, and the provision of recreational spaces. Coastal forests, including those along the Tanzanian coastline, play a crucial role in mitigating coastal erosion and sustaining fisheries (Kilahama, 2021).

Rapid urbanization in East Africa presents significant challenges. The expansion of cities into peri-urban forested regions results in habitat loss and heightened human-wildlife conflicts. The Green Belt Movement in Kenya exemplifies the capacity of community-led conservation initiatives to tackle these challenges.

1.3.5 Tanzania: A Country of Rich Forest Heritage

Tanzania's forests encompass approximately 48.1 million hectares, featuring a variety of ecosystems, including miombo woodlands, mangroves, and coastal forests (MNRT, 2022). The forests are essential to Tanzania's socio-economic development, underpinning agriculture, energy, and tourism.

Tanzanian forests play a crucial role in regulating water cycles, preventing soil erosion, and providing habitat for endemic species (Mgaya & Samwel, 2021). They provide socio-economic support to millions of Tanzanians reliant on forests for fuelwood, construction materials, and traditional medicine. Urban forests in cities such as Dar es Salaam provide benefits including the reduction of heat stress and the availability of green spaces for recreational activities (Nyika, 2020).

The Tanzanian government has implemented policies, including the National Forest Policy of 1998, to reconcile conservation and development, with a focus on participatory forest management (MNRT, 2022). Challenges such as illegal logging, encroachment, and limited funding continue to exist.

1.3.6 Kazimzumbwi Forest Reserve: A Case Study of Ecological and Socio-economic Integrities

Kazimzumbwi Forest Reserve, situated near Dar es Salaam, illustrates the dual function of forests in promoting ecological and socio-economic sustainability. The coastal forests of Tanzania represent a significant biodiversity hotspot and carbon sink, essential for mitigating urban heat and the impacts of climate change (Mwakalila et al., 2023).

Ecological Integrities

The forest hosts numerous endemic and threatened species, thereby enhancing Tanzania's biodiversity (MNRT, 2022). It facilitates hydrological processes through groundwater recharge and local temperature regulation. The role of carbon sequestration is particularly significant given the rapid urbanization of Dar es Salaam, which has led to increased greenhouse gas emissions.

Socio-economic Integrities

Kazimzumbwi Forest supplies vital resources to adjacent communities, such as firewood, honey, and medicinal plants (Mgaya & Samwel, 2021). The potential for ecotourism exists, which may generate income for local populations and support conservation initiatives. The forest enhances urban resilience by reducing flooding and serving as a buffer against land degradation.

Challenges and Opportunities

Kazimzumbwi, despite its significance, encounters challenges including illegal logging, agricultural encroachment, and insufficient management. Community-based conservation programs, with support from local and international stakeholders, are crucial for maintaining ecological integrity (Nyika, 2020).

Kazimzumbwi's example highlights the necessity of integrated urban planning that recognizes forests as essential elements of urban resilience and sustainability, rather than merely as natural assets.

CHAPTER TWO

2. LAND USE AND LAND COVER CHANGES (LULCC)

2.1 Dynamics of Urbanization and Forest Cover

Urbanization represents a significant transformation in contemporary society, marked by the growth of urban areas and changes in land use and land cover (LULC). Urbanization stimulates economic growth and societal advancement; however, it presents considerable challenges to forest ecosystems, especially in urban and peri-urban regions. This section analyzes the dynamics of urbanization and its effects on forest cover globally, with a specific emphasis on Africa, Sub-Saharan Africa, East Africa, Tanzania, and the Kazimzumbwi Forest Reserve.

2.1.1 Global Perspective: Urbanization and LULC Dynamics

Urban areas are experiencing rapid expansion, with projections indicating that cities will accommodate nearly 70% of the global population by 2050 (United Nations, 2022). This growth is associated with significant alterations in land use and land cover, frequently marked by the transformation of forests, agricultural land, and wetlands into urbanized regions. From 2000 to 2020, approximately 100 million hectares of forest were lost globally, primarily attributed to urban expansion and infrastructure development (FAO, 2022).

The processes associated with urbanization frequently lead to fragmented forest landscapes, diminished biodiversity, and compromised ecosystem services. Tropical forests in South America and Southeast Asia have undergone considerable deforestation as a result of urban sprawl and industrial agriculture (Hansen et al., 2021). Urban forests and green spaces are essential for preserving ecological integrity in urban areas, yet they are often overlooked due to competing land-use pressures.

Mitigation efforts encompass the adoption of sustainable urban planning practices, the promotion of urban reforestation, and the integration of green infrastructure within cityscapes. Urban greening initiatives in cities like Singapore and Vancouver exemplify the capacity to harmonize urban development with forest conservation (Nowak et al., 2023).

2.1.2 Africa: Urbanization's Impact on Forests

Africa is experiencing rapid urbanization, with projections indicating that urban populations will triple by 2050 (UN-Habitat, 2022). The rapid urbanization places substantial pressure on forest ecosystems, resulting in notable land use and land cover changes. Forests are being cleared to facilitate the growth of urban areas, and peri-urban forests are increasingly encroached upon for agricultural, residential, and infrastructural purposes (FAO, 2023).

Urbanization in Africa, alongside deforestation, exacerbates forest degradation by heightening the demand for fuelwood and charcoal, especially in low-income urban regions (African Forest Forum, 2022). Cities such as Kinshasa, Lagos, and Nairobi have experienced significant declines in peri-urban forest cover in the last thirty years.

Urbanization presents both challenges and opportunities for sustainable forest management. The African Forest Landscape Restoration Initiative (AFR100) seeks to restore 100 million hectares of degraded land across Africa by 2030, highlighting the necessity of incorporating forests into urban planning frameworks to address the effects of urbanization (UNEP, 2022). Examples of success, such as the restoration of Nairobi's Karura Forest, illustrate the effectiveness of community engagement and policy support in mitigating land use and land cover changes.

2.1.3 Sub-Saharan Africa: Land-Use Transitions and Forests

Sub-Saharan Africa (SSA) is undergoing notable land-use transitions influenced by urbanization, population growth, and economic development. Urban sprawl significantly contributes to land use and land cover changes, leading to a concerning reduction in forest cover. From 2001 to 2021, SSA experienced a loss of more than 14 million hectares of tree cover, predominantly in peri-urban regions (Global Forest Watch, 2022).

Urbanization in Sub-Saharan Africa frequently transpires in an unstructured and informal way, resulting in fragmented landscapes and diminished ecological functions. Cities such as Addis Ababa, Ethiopia, and Kampala, Uganda, have undergone significant deforestation in peri-urban regions as a result of informal settlements and agricultural expansion (Myers et al., 2023).

Despite these challenges, emerging opportunities for sustainable land use and land cover management exist. Urban forestry programs, agroforestry in peri-urban areas, and the creation of green corridors are increasingly recognized as effective strategies for reconciling urban development with forest conservation (African Development

Bank [AfDB], 2022). These approaches seek to restore ecological balance while addressing socio-economic requirements.

2.1.4 East Africa: Forest Cover and Urban Expansion

The forest ecosystems of East Africa, including the Eastern Arc Mountains and coastal forests, are significantly susceptible to land use and land cover changes resulting from urbanization. Urban centers such as Nairobi, Kampala, and Dar es Salaam have experienced significant growth, leading to the transformation of forested regions into residential and commercial developments (Wanjiru & Matsui, 2022).

Urbanization in Tanzania has notably affected forest cover, with annual deforestation rates approximated at 1% (MNRT, 2022). Peri-urban forests, serving as ecological buffers, are notably impacted. The Pugu and Kazimzumbwi Forest Reserves near Dar es Salaam have experienced significant degradation as a result of encroachment and illegal logging (Mwakalila et al., 2023).

Efforts to address these dynamics involve the implementation of Tanzania's National Forest Policy and the promotion of urban greening initiatives. Initiatives like the Dar es Salaam Greenbelt Initiative focus on the restoration of degraded forests and the incorporation of green infrastructure within urban environments (Nyika, 2020).

2.1.5 Tanzania: Urbanization and Forest Dynamics

Tanzania exhibits an urbanization rate of 4.7% annually, one of the highest in East Africa, resulting in notable land use and land cover changes (National Bureau of Statistics [NBS], 2023). Urban expansion has led to the depletion of natural forests, especially in peri-urban regions, where forest resources are significantly utilized for fuelwood and construction materials (Mgaya & Samwel, 2021).

Urban centers like Dar es Salaam and Dodoma have experienced significant declines in forest cover as a result of infrastructure development and population growth. Initiatives such as Tanzania's Urban Resilience Program highlight the importance of incorporating forests into urban planning to enhance climate adaptation and mitigate disaster risks (MNRT, 2022).

2.1.6 Kazimzumbwi Forest Reserve: A case study of LULC dynamics

The Kazimzumbwi Forest Reserve, located near Dar es Salaam, provides a stark example of the interplay between urbanization and forest cover. Established in 1936, this coastal forest has been severely impacted by LULC changes driven by urban sprawl, illegal activities, and weak enforcement of conservation policies (Mwakalila et al., 2023).

Land Use Land Cover (LULC) Changes

Data collection

This study integrates satellite remote sensing data with ground-based surveys to evaluate the current Land Use and Land Cover (LULC) status of the Kazimzumbwi Forest Reserve in Tanzania. Combining these methods enhances the accuracy of LULC classification and provides a comprehensive understanding of land use dynamics within the reserve (Regasa & Nones, 2022).

Satellite data acquisition

Multispectral satellite imagery indicated in Table 2.1 was obtained from the United States Geological Survey (USGS) Earth Explorer platform (https://earthexplorer.usgs.gov/). Specifically, Landsat imagery was selected due to its long-term data availability, appropriate spatial resolution, and cost-effectiveness for environmental monitoring (Giri, 2012). The study utilized data from Landsat 4–5 Thematic Mapper (TM), Landsat 7 Enhanced Thematic Mapper Plus (ETM+), and Landsat 8 Operational Land Imager/Thermal Infrared Sensor (OLI/TIRS). Images were acquired for the dry season months of June and July, corresponding to minimal cloud cover periods in the region. Only images with less than 20% cloud cover were selected to ensure data quality and reliability (Mansour et al., 2023).

Table 2.1: The Metadata on the satellite images acquired

Satellite	Sensor	Year	Acquisition date	Spatial resolution
Landsat 8	OLI/TIRS	2024	June 20	30m
Landsat 7	ETM +	2004	June 24	30m
Landsat 4-5	TM	1994	July 03	30m

Spectral bands and spatial resolution

The spectral bands used for image classification included Blue (Band 2), Green (Band 3), Red (Band 4), Near-Infrared (NIR) (Band 5), Short-Wave Infrared 1 (SWIR 1) (Band 6), and Short-Wave Infrared 2 (SWIR 2) (Band 7). These bands are crucial for differentiating various land cover types due to their sensitivity to vegetation health, soil moisture, and other surface characteristics (Gyamfi-Ampadu et al., 2020). The spatial resolution of 30 meters provided by these bands is adequate for monitoring forest dynamics and detecting land cover changes at a local scale (Massey et al., 2023).

Ground truth data collection

Field surveys were conducted using handheld Global Positioning System (GPS) devices to collect ground truth data. These data points were strategically distributed throughout the Kazimzumbwi Forest Reserve to represent the diversity of land cover types, including dense forest, degraded forest, agriculture, grassland, and built-up areas (Congalton & Green, 2019). The ground truth data served two primary purposes namely training the classification algorithm, and validation. Training the classification algorithm by providing sample data to train the remote sensing classification algorithms for accurate land cover identification; and validation by assessing the accuracy of the classified images by comparing them with actual field conditions (Regasa & Nones, 2022).

Data preprocessing

Preprocessing steps were essential to ensure the reliability of the satellite data. These steps included:

i. Radiometric calibration: Correcting sensor errors and normalizing the images to reflect true surface conditions (Chander et al., 2009).
ii. Atmospheric correction: Removing atmospheric effects using the Landsat Ecosystem Disturbance Adaptive Processing System (LEDAPS) to retrieve surface reflectance values (Vermote et al., 2016).
iii. Geometric correction: Aligning images to a common coordinate system to ensure spatial accuracy (Jensen, 2015).

These preprocessing steps are critical for comparing images over time and accurately detecting land cover changes (Song et al., 2001).

Data integration and analysis

The preprocessed satellite images and ground truth data were integrated into a Geographic Information System (GIS) for analysis. The following methods were employed:

i. Supervised classification: Utilizing algorithms like Maximum Likelihood Classification to categorize land cover types based on spectral signatures (Lillesand et al., 2015).

ii. Accuracy assessment: Comparing classified images with ground truth data to calculate metrics such as Overall Accuracy and Kappa Coefficient (Congalton & Green, 2019).

iii. Change detection analysis: Identifying and quantifying land cover changes over the study period to understand trends and patterns (Lu et al., 2004).

Importance of the selected methods

i. Integration of satellite and ground data: Enhances classification accuracy and provides a robust dataset for analysis (Regasa & Nones, 2022).

ii. Use of landsat imagery: Offers a balance between spatial resolution and temporal coverage, suitable for long-term environmental monitoring (Giri, 2012).

iii. Selection of spectral bands: The chosen bands are effective in distinguishing vegetation types and detecting changes in forest cover (Gyamfi-Ampadu et al., 2020).

Data analysis and processing

Image processing and classification

This study employed Landsat satellite imagery for analyzing Land Use and Land Cover (LULC) dynamics in the Kazimzumbwi Forest Reserve. The Landsat images used were Level 1 terrain-corrected data, pre-processed to address geometric and radiometric distortions, ensuring consistency and reliability for classification (Pinto et al., 2020). As the images were acquired on different dates and using various sensors, a harmonization process was conducted to ensure compatibility for a multitemporal analysis. High-resolution images from Google Earth (2024) were utilized as reference data for visualizing target land cover types (Zhao & Zhu, 2022).

To classify LULC, supervised classification using the Random Forest (RF) algorithm was implemented. RF, developed by Breiman (2001), is a robust non-parametric classification technique suitable for handling complex datasets and temporal analyses (Wale et al., 2023). The algorithm combines multiple decision trees with a majority voting rule to produce accurate classifications. Key advantages include its ability to handle large datasets and its resistance to overfitting, making it ideal for this study's multitemporal LULC analysis (Latella et al., 2021).

The classification was based on training datasets derived from field survey points and high-resolution satellite imagery. The images were projected into the WGS 84/UTM Zone 37 South coordinate system to ensure geospatial consistency. The predefined land cover classes included closed forests, open forest/woodland, grassland, bare land, shrubs, and built-up areas, reflecting the diversity of land cover types in the Kazimzumbwi Forest Reserve. Validation of the classification was conducted using one-third of the ground survey points collected in the field, with the classification process applied to all selected time periods.

LULC types	Description
Closed forest	Area with enclosed formations, tall trees and enclosed formations of dense forest
Open forest	with mixed forest formations, open stands and whose tree tops are less joined or form a clear canopy
Shrubs	woody perennial plants, generally of more than 0.5 m and less than 5 m in height on maturity and without a definite crown
Grassland	Ground covered by vegetation dominated by grasses, with little or no tree cover
Bare land	Areas with no dominant vegetation cover on at least 90 % of the area
Built-up areas	Areas which are characterized by an artificial cover

Accuracy assessment

Accuracy assessment is a crucial component of LULC classification, ensuring that the outputs represent actual ground conditions. Following the methodology proposed by Bunyangha et al. (2021), this study implemented a comprehensive accuracy evaluation to validate the classification results. High-resolution imagery from Google Earth (2024) was used as a reference for verifying the spectral and spatial accuracy of the classified maps (Leta et al., 2021).

The assessment included calculating metrics such as Overall Accuracy and the Kappa Coefficient, a statistical measure of agreement between classified data and reference ground truth points. The Kappa Coefficient, calculated using the formula by Congalton (1991), ranges from 0 to 1, with values closer to 1 indicating high classification accuracy and values near 0 suggesting minimal agreement (Rwanga et al., 2017). The metrics were computed by comparing the correctly classified pixels to the total number of reference points, ensuring reliable validation (Islami et al., 2022).

Land use and land cover change detection

Change detection for LULC was performed using the Post Classification Comparison (PCC) method, which compares classified images from different time periods. This approach overlays classifications from two dates to generate a transition matrix that quantifies changes in land cover categories over time (Kalaiyarasi et al., 2023). The rows of the matrix represent the land use types at the initial date (t_1), while the columns represent land use types at the subsequent date (t_2). Diagonal elements indicate areas of no change, while off-diagonal elements capture conversions between categories.

For this study, the PCC method was applied to assess transitions across the six predefined LULC categories. A 6 × 6 transition matrix quantified the changes in area for each class, allowing for the detailed analysis of conversions such as forest degradation, urban expansion, or agricultural encroachment. This technique is particularly effective for understanding the dynamics of LULC change in ecologically sensitive areas like forest reserves (Pinto et al., 2020).

Annual rate of land use and land cover change

The annual rate of LULC change was calculated using the formula proposed by Puyravaud (2003):

$$r = \frac{\ln(C2) - \ln(C1)}{t2 - t1}$$

Where:

C_1 and C_2 are the areas of a given land use class at times t_1 and t_2, respectively, r represents the annual rate of change.

A positive rate indicates an increase in the specific LULC class, a negative rate indicates a decrease, and a value near zero suggests stability (Liu et al., 2020). This analysis provides insights into the rates at which different land cover types are changing, highlighting critical areas of concern, such as deforestation or urban expansion. Quantifying these changes enables targeted conservation interventions for the Kazimzumbwi Forest Reserve.

Future prediction of land use and land cover (LULC)

The study employed the TerrSet Geospatial Monitoring and Modeling System, specifically the Land Change Modeler (LCM), to forecast future Land Use and Land Cover (LULC) patterns for the years 2034 and 2054. This software is recognized for its robust dynamic projection capabilities, extensive calibration, and ability to simulate diverse land cover types effectively (Eastman, 2009; Leta et al., 2021). Using historical satellite images classified during this research, the LCM model analyzed land use changes between two dates t_1 (initial date) and t_2 (final date). These analyses were then extrapolated to predict LULC changes for t_3 (2034) and t_4 (2054), utilizing transition potential maps generated from the historical data (Qacami et al., 2023).

Modeling techniques and transition potential maps

The LCM model utilizes transition potential maps to model land use and land cover changes. Transition potential maps were produced using three principal methodologies: (i) Multi-Layer Perceptron (MLP) Neural Network, (ii) Logistic

Regression, and (iii) Machine Learning using Similarity-Weighted Instances (SimWeight). The MLP neural network method was chosen for its low parameter demands, intuitive interface, and enhanced automation capabilities (Eastman, 2016; Regasa & Nones, 2022). The MLP neural network architecture comprises three fundamental layers:

(i) Input Layer: Comprises the driving variables affecting LULC alterations, including proximity to disturbances, urban regions, waterways, and thoroughfares.

(ii) Hidden Layer: Consists of computational nodes that analyze input data and compute transition probabilities.

(iii) Output Layer: Generates transition potential maps that signify the probability of a certain land cover type transforming into another.

The model produces a transition sub-model for each land cover transition. When numerous transitions are influenced by analogous factors, these sub-models are consolidated to provide a composite transition adequacy map, illustrating the susceptibility and probability of transitions for particular land cover categories (Bunyangha et al., 2021).

Driving variables for LULC change predictions

Precise future forecasts necessitate access to historical land use and land cover (LULC) maps and a compilation of factors driving changes in land cover. This study examined the following dynamic driving variables: distance from disturbances (e.g., agricultural activities or deforestation), closeness to streams, distance to metropolitan areas, and proximity to roadways. The dynamic variables were augmented with static variables, including elevation and slope, resulting in a full dataset for modeling transitions (Regasa & Nones, 2021, 2022). The driving variables were essential for generating transition potential maps and modeling future land use and land cover scenarios.

Analysis of historical LULC trends

Historical land use and land cover maps from 1994, 2004, and 2024 were utilized to identify trends and guide forecasts for 2034 and 2054. A business-as-usual (BAU) scenario was implemented, presuming that future land use alterations will persist in accordance with historical trends (Clerici et al., 2019). The BAU scenario establishes a benchmark for assessing the potential effects of ongoing development, urbanization, and deforestation on the Kazimzumbwi Forest Reserve.

Forecasting future LULC scenarios

The LCM model incorporated two sophisticated methodologies to forecast future trends:

Multiple Objective Land Allocation (MOLA): A method employed for the geographic optimization of land cover distribution while satisfying established objectives (Eastman, 2009). Cellular Automata (CA) which is a geographic modeling methodology that represents land cover changes predicated on the existing state and adjacent conditions (Eastman, 2016). These techniques were employed to spatially allocate Markov transitions, measuring the probability of land cover alterations across time. The Markov model posits that future land use and land cover states are contingent upon the present state and historical transition probabilities. Transition matrices were created to ascertain the odds of one LULC type changing to another, enabling projections of spatial patterns for 2034 and 2054 (Bunyangha et al., 2021).

Expected future LULC trends

The LULC projections indicate that forthcoming alterations in the Kazimzumbwi Forest Reserve will be shaped by analogous factors as past patterns, including (i) deforestation and agricultural development result in diminished forest cover and an increase in agricultural and barren land, (ii) urbanization which facilitating the proliferation of developed regions, especially adjacent to primary thoroughfares and metropolitan hubs, and (iii) conservation initiatives as possibly mitigating the degradation rate in protected areas (Regasa & Nones, 2022). The study highlights the necessity of proactive land management and legislative measures to alleviate negative effects on the forest ecosystem.

Results of LULC Classification for Kazimzumbwi Forest Reserve from 1994 to 2024

According to the LULC categorization data that is presented in Tables 2.2-2.4 and Figures 2.1-2.3 for the years 1994, 2004, and 2024 for Kazimzumbwi Forest Reserve, there have been significant variations in the types of land cover that have occurred over the course of the past thirty years. Both Congalton (1991) and Bunyangha et al. (2024) discovered that there was a significant agreement between the categorized maps and the ground truth data, which indicates that the categorization reached a high level of accuracy. Additional evidence that this discovery is correct is provided by kappa values of 0.85 (1994), 0.87 (2004), and 0.91 (2024).

Table 2.2: LULC composition and change trend of Kazimzumbwi from 1994 to 2024

LULC type	1994		2004		2024		change detected (ha)		Rate of change per annum (%)	
	ha	%	ha	%	ha	%	1994-2004	2004-2024	1994-2004	2004-2024
Closed forest	2722.28	54.7	1577.58	31.7	886.64	17.8	-1144.70	-690.94	-5.31	-2.84
Open woodland	1399.65	28.1	601.068	12.1	1390.09	27.9	-798.58	789.02	-8.11	4.28
Shrubs	372.28	7.5	791.796	15.9	1160.98	23.3	419.51	369.18	7.84	1.93
Grassland	235.89	4.7	1666.42	33.5	925.22	18.6	1430.53	-741.21	21.59	-2.9
Bare land	245.07	4.9	338.298	6.8	540.01	10.9	93.23	201.71	3.28	2.37
Built-up	0.00	0.0	0		72.23	1.5	0	72.23		
Total	4975.17		4975.17		4975.17					

Table 2.3: LULC transition matrix 1994 to 2004, Area in ha

LULC types	Closed forest	Open woodland	Shrubs	Grassland	Bare land	Built-up	Total 1994
Closed forest	1289.94	412.15	407.75	497.05	115.39	0	2722.28
Open woodland	209.27	134.78	249.65	692.23	113.97	0	1399.65
Shrubs		27.40	113.20	190.79	41.09	0	372.28
Grassland	61.06	9.05	21.19	144.40	0	0	235.89
Bare land	17.32	17.70	0	141.95	68.10	0	245.07
Built-up	0	0	0	0	0	0	0
Total 2004	1577.58	601.07	791.80	1666.42	338.29		

Table 2.4: LULC transition matrix 2004 to 2024, Area in ha

LULC types	Closed forest	Open woodland	Shrubs	Grassland	Bare land	Built-up	Total 2004
Closed forest	615.99	424.68	97.04	193.90	245.97	0	1577.58
Open woodland	77.79	294.04	0	155.06	64.87	9.30	601.07
Shrubs	111.78	0	455.70	143.24	67.97	13.05	791.80
Grassland	79.79	611.02	490.13	355.87	87.81	41.61	1666.42
Bare land	1.29	60.35	118.11	77.14	73.40	8.27	338.30
Built-up	0	0	0	0	0	0	0
Total 2024	886.64	1390.09	1160.98	925.22	540.01	72.23	

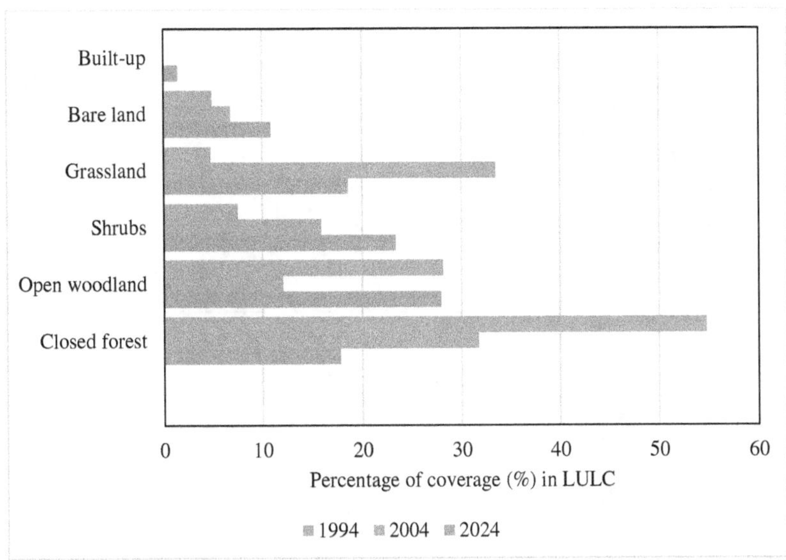

Figure 2.1: Trend of LULC in Kazimzumbwi Forest Reserve from 1994 to 2024

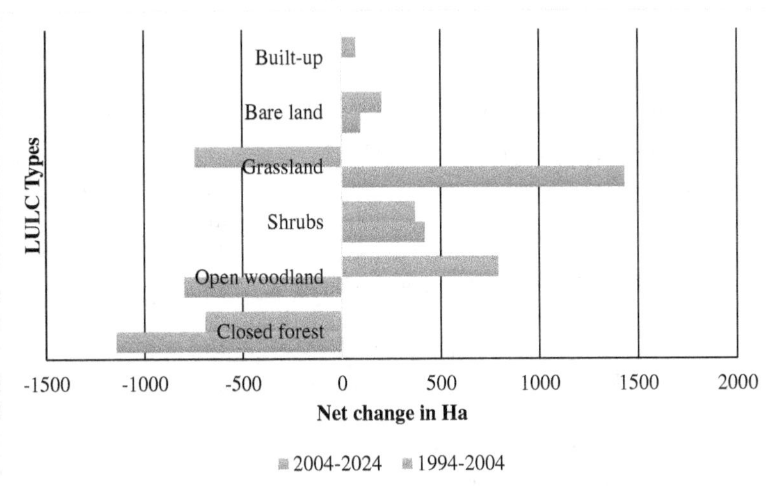

Figure 2.2: Net change of LULC for Kazimzumbwi Forest Reserve from 1994 to 2024

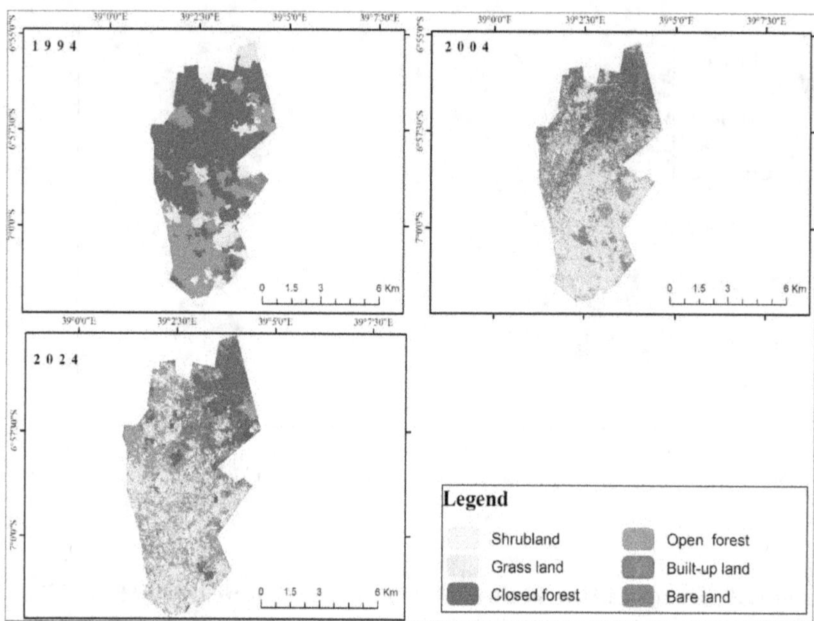

Figure 2.3: Classified LULC maps of Kazimzumbwi Forest Reserve from 1994 to 2024

LULC Classification Results of Kazimzumbwi Forest Reserve for 1994

In 1994, the closed forest covered an area of 2722.28 hectares, representing 54.7% of the total expanse of 4975.17 hectares within the Kazimzumbwi Forest Reserve (Table 3, Figure 1). The open forest spans 1399.65 hectares, constituting 28.1% of the total area, while shrubs cover 372.28 hectares, representing 7.5% of the area. The expanse of bareland constituted the least significant portion, encompassing 245.07 hectares (4.9%), with no recorded instances of built-up areas during this timeframe. The dominance of closed forest signifies the reserve's unspoiled state in the early 1990s, highlighting its vital role in maintaining ecological integrity and providing essential ecosystem services, such as biodiversity conservation and carbon sequestration (Chapin et al., 2020). Comparable observations were documented in the Nairobi Arboretum by Nduati and Kamau (2020), revealing that closed forest once encompassed 58% of the region in the 1990s, underscoring the ecological significance of urban and peri-urban forests in mitigating environmental degradation.

LULC Changes of Kazimzumbwi Forest Reserve in 2004

By 2004, the closed forest area in Kazimzumbwi had diminished to 31.7%, indicating a decline of approximately 23% over the previous decade. The increase in open forest and shrub areas is linked to forest degradation, mainly resulting from illegal logging, agricultural encroachment, and shifting cultivation practices. The increasing prevalence of open forest and shrubland signifies a transitional phase of land cover degradation, consistent with Clerici et al. (2019), who identified forest loss in Colombian urban forests associated with similar anthropogenic pressures. The persistent occurrence of bare land at around 5% suggests that, despite evident deforestation, the forest reserve policies may have reduced the total land clearing. This trend aligns with the findings of Mansour et al. (2023), who observed that strict enforcement of protected area policies in arid and semi-arid regions decreased the probability of complete land transformation.

LULC Trends of Kazimzumbwi Forest Reserve in 2024

By 2024, the percentage of closed forest has decreased to 17.8% of the total area, reflecting continued and accelerated degradation of the forest reserve. This reflects a decline of over 67% in closed forest cover since 1994, demonstrating the cumulative impacts of deforestation, agricultural expansion, and urban encroachment. The landscape was primarily defined by open forest and shrubs, with built-up areas, previously nonexistent, now accounting for 1.5% of the total area. The development of areas within the reserve signifies illegal encroachment, contravening Tanzanian policies and regulations regarding protected forests. The Forest Act of Tanzania (2002) prohibits settlement and construction in designated forest reserves. This finding is consistent with research from Ethiopia, where Yirdaw et al. (2020) identified similar encroachment trends in protected forest areas, linked to rapid urbanization and insufficient enforcement mechanisms.

The patterns observed in Kazimzumbwi align with global trends in urban and peri-urban forests, as evidenced by related studies. For example:

(i) Nairobi Arboretum, Kenya: Nduati and Kamau (2020) reported a 25% reduction in closed forest cover from 1995 to 2015, linked to urban expansion and informal settlements. Similar to Kazimzumbwi, developed areas emerged within the arboretum despite legal protections.

(ii) Colombian Urban Forests: Clerici et al. (2019) documented a significant reduction in urban forests in Colombia, indicating a 20% decrease in closed forest cover over a thirty-year span. Illegal logging and agricultural encroachment are recognized as key drivers, akin to the pressures faced by Kazimzumbwi.

(iii) Yirdaw et al. (2020) reported that encroachment into protected forests in Ethiopia resulted in a 15% increase in shrub and open forest cover, as closed forests transitioned to degraded states. This indicates the trajectory observed in Kazimzumbwi.

The significant decline in closed forest cover in Kazimzumbwi has considerable ecological and socio-economic implications. The reduced capacity of the forest to sequester carbon, maintain biodiversity, and regulate climate undermines its role as an ecological buffer in the peri-urban landscape of Dar es Salaam. The development of built-up areas within the reserve poses a risk to hydrological cycles and may intensify soil erosion, consequently undermining the integrity of the forest (Chapin et al., 2020). The socio-economic decline of Kazimzumbwi demonstrates a misalignment between conservation goals and the livelihood needs of the local community. Gyamfi-Ampadu et al. (2020) documented similar results in Ghanaian urban forests, emphasizing the need for integrated land management strategies to balance ecological conservation with socio-economic development.

The findings from Kazimzumbwi Forest Reserve for the period 1994-2024 highlight the need for effective policy enforcement, community engagement, and sustainable land use planning to protect urban and peri-urban forests. Analysis of these trends in relation to comparable global studies indicates that the challenges faced in Kazimzumbwi are not isolated but reflect a broader pattern of forest degradation in areas undergoing rapid urbanization.

Results of forecasted LULC Classification for Kazimzumbwi Forest Reserve from 2024 to 2054

LULC dynamics under the business-as-usual (BAU) scenario

The projections for Land Use and Land Cover (LULC) in the Kazimzumbwi Forest Reserve from 2024 to 2054 under the Business-as-Usual (BAU) scenario, as illustrated in Tables 2.5-2.6 and Figures 2.4-2.5, indicate notable changes in land cover types. The observed changes underscore continuous ecological transformations shaped by human activities and natural processes. The results demonstrate a significant reduction in closed forest cover, alongside a rise in open woodland and grassland areas. In contrast, shrubs, bare land, and built-up areas demonstrate differing levels of reduction during the same timeframe.

Table 2.5: LULC composition and change trend of Kazimzumbwi from 2024 to 2054

LULC type	2024		2034		2054		2024-2034		2034-2054		2024-2054	
	Ha	%	ha	%	ha	%	ha	%	ha	%	ha	%
Closed forest	886.64	17.8	595.24	12.0	537.58	10.8	-291.40	-32.9	-57.66	-9.7	-349.06	-39.4
Open woodland	1390.09	27.9	1748.03	35.1	1625.97	32.7	357.94	25.7	-122.06	-7.0	235.88	17.0
Shrubs	1160.98	23.3	1225.13	24.6	1131.29	22.7	64.15	5.5	-93.83	-7.7	-29.69	-2.6
Grassland	925.22	171.3	876.83	17.6	1098.81	22.1	-48.38	-5.2	221.98	25.3	173.59	18.8
Bare land	540.01	10.9	464.26	9.3	516.66	10.4	-75.75	-14.0	52.40	11.3	-23.35	-4.3
Built-up	72.23	1.5	65.68	1.3	64.86	1.3	-6.55	-9.1	-0.82	-1.2	-7.37	-10.2
Total	**4975.17**		**4975.17**		**4975.17**							

Table 2.6: LULC transition matrix from 2024 to 2054, Area in ha

LULC types	Closed forest	Open forest	Shrub land	Grassland	Bare land	Built-up	Total 2024
Closed forest	373.02	228.02	3.99	163.82	116.69	1.10	**886.64**
Open forest	7.43	728.03	349.43	300.82	3.53	0.84	**1390.09**
Shrubland	147.264	335.37	401.97	270.08	5.15	1.14	**1160.98**
Grassland	4.39	310.68	237.83	359.26	12.45	0.62	**925.22**
Bare land	4.76	21.38	136.35	2.55	378.52	0.34	**543.90**
Built-up	0.71	2.49	1.73	2.28	0.31	60.82	**68.34**
Total 2054	**537.58**	**1625.97**	**1131.29**	**1098.81**	**516.66**	**64.86**	**4975.17**

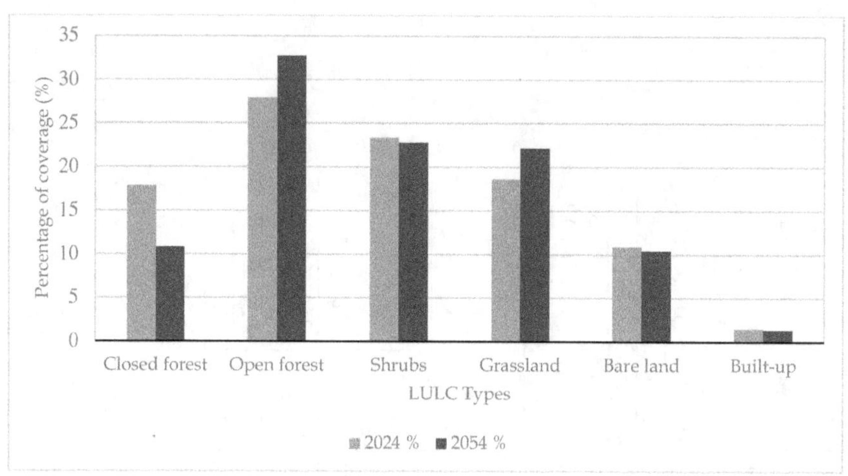

Figure 2.4: Trend of LULC in Kazimzumbwi Forest Reserve from 2024 to 2054

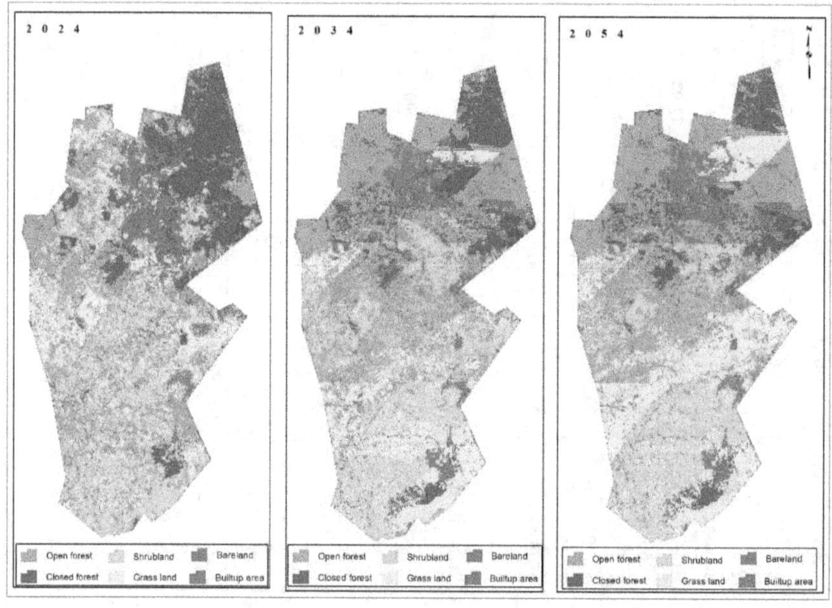

Figure 2.5: Future trends in LULC of Kazimzumbwi Forest Reserve from 2024 to 2054

Decline in closed forest

The closed forest area is expected to decrease by 39.4%, from 886.64 hectares in 2024 to 537.58 hectares in 2054, indicating a substantial loss of an essential ecological resource. The decline represents a reduction of approximately 349.06 hectares over the 30-year period, underscoring the ongoing issues of deforestation, illegal logging, and agricultural encroachment within the reserve.

This pattern aligns with global findings, exemplified by the Nairobi Arboretum, where Nduati and Kamau (2020) documented a 25% decrease in closed forest cover from 1995 to 2015, attributed to urban expansion and unsustainable resource extraction. Clerici et al. (2019) reported reductions in forest cover within Colombian urban forests, linking these losses to unregulated human activities and inadequate enforcement of conservation policies.

Increase in open woodland and grassland

Open woodland and grassland are expected to rise by 17% and 18.8%, respectively. The expansion results from the degradation of closed forest areas into more open and fragmented landscapes. Although these land cover types possess some ecological value, their effectiveness in delivering essential ecosystem services like carbon sequestration, water regulation, and biodiversity support is inferior to that of closed forests (Chapin et al., 2020).

Similar trends have been noted in Ethiopian protected forests, with Yirdaw et al. (2020) indicating that closed forests frequently transition into open woodlands and grasslands as a result of heightened human activities, including grazing and fuelwood collection. The transformation of dense forest into less dense land cover types presents challenges for preserving ecological integrity in urban and peri-urban environments.

Reduction in shrubs, bare land, and built-up areas

Shrubs, bare land, and built-up areas are expected to decline by 2.6%, 4.3%, and 10.2%, respectively, during the same timeframe. The decrease in shrubs and bare land may be associated with the natural transition of these regions into grasslands or woodlands, influenced by secondary succession or alterations in land management practices (Bunyangha et al., 2021). The observed decline in built-up areas is a typical and may indicate methodological adjustments or reclassification of land use during the modeling process. This finding diverges from global trends, wherein urban expansion typically results in an increase in built-up areas, especially near forest reserves (Mansour et al., 2023).

Ecological and socio-economic implications

The projected decline in closed forest cover and concurrent increase in open woodland and grassland carry profound ecological and socio-economic implications:

Loss of ecosystem services	Closed forests play a vital role in carbon sequestration, the preservation of biodiversity, and the regulation of microclimates. Their reduction decreases the forest's ability to mitigate climate change and sustain ecological stability (Chapin et al., 2020)
Decreased biodiversity	Habitat fragmentation resulting from forest degradation negatively impacts biodiversity. Species dependent on dense forest cover may experience habitat loss and heightened extinction risk (Gyamfi-Ampadu et al., 2020).
Socio-economic challenges	The degradation of Kazimzumbwi Forest Reserve adversely affects the livelihoods of communities reliant on forest resources, including fuelwood, medicinal plants, and various non-timber forest products (NTFPs). Comparable results were observed in urban forests of Ghana, indicating that forest degradation had a disproportionate impact on low-income communities dependent on forest resources (Gyamfi-Ampadu et al., 2020)
Policy and management gaps	The ongoing issues of deforestation and land-use change underscore the necessity for enhanced enforcement of conservation policies and the implementation of community-based forest management strategies. The Forest Act of Tanzania (2002) establishes a framework for sustainable forest management; however, deficiencies in implementation and enforcement persist, compromising its effectiveness (Moshi & Kweka, 2024).

Comparison with similar studies

The results from Kazimzumbwi Forest Reserve correspond with global patterns identified in urban and peri-urban forests:

(i) Nduati and Kamau (2020) documented a transition from closed forests to open woodlands and grasslands in the Nairobi Arboretum, Kenya, attributed to deforestation and urban pressures.

(ii) Colombian Urban Forests: Clerici et al. (2019) emphasized the gradual degradation of closed forests into open land cover types, driven by inadequate conservation enforcement and socio-economic pressures.

(iii) Ethiopian Protected Forests: Yirdaw et al. (2020) examined the transformation of dense forest regions into open and fragmented landscapes, highlighting the impact of human activities on land cover change.

The BAU scenario for Kazimzumbwi Forest Reserve highlights the necessity for sustainable land management practices and enhanced policy enforcement to address forest degradation. Proactive interventions, including reforestation, community engagement, and enhanced monitoring systems, are vital for maintaining the ecological and socio-economic health of this important peri-urban forest.

2.2 Drivers of Land Use and Land Cover Change in Urban Areas

Land Use and Land Cover Change (LULCC) in urban areas represents a dynamic process influenced by socio-economic, demographic, political, and environmental factors. Urban LULCC, frequently characterized by the transformation of natural landscapes into constructed environments, presents considerable challenges to ecological integrity and socio-economic resilience. This section analyzes the factors influencing LULCC from a global perspective down to specific contexts, including Africa, Sub-Saharan Africa, East Africa, Tanzania, and the Kazimzumbwi Forest Reserve.

2.2.1 Global drivers of LULCC in urban areas

Globally, rapid urbanization and population growth serve as primary drivers of land use and land cover change (LULCC). The global urban population increased from 751 million in 1950 to over 4.4 billion in 2021, resulting in heightened demand for housing, infrastructure, and industrial facilities (United Nations, 2022). This demand propels the growth of urban regions, frequently compromising forests, wetlands, and agricultural lands.

Economic development serves as a crucial catalyst. Industrialization and urban expansion frequently result in the transformation of peri-urban forests into industrial and residential zones (FAO, 2023). Technological advancements can enhance economic growth but may also drive land use and land cover change (LULCC) by enabling the exploitation of natural resources, exemplified by mechanized logging and large-scale construction projects (Hansen et al., 2021).

Policy and governance are essential components. Insufficient land management policies, poor enforcement of conservation regulations, and corruption frequently exacerbate land use and land cover change (LULCC). Poor urban planning in developing countries often results in unregulated land conversion and the encroachment of informal settlements into natural ecosystems (Seto et al., 2012).

Climate change serves as both a catalyst and an outcome of land use and land cover change (LULCC). Increasing temperatures and altered precipitation patterns can impair forests, rendering them more vulnerable to conversion for urban or agricultural purposes. Urban expansion concurrently intensifies climate change through heightened greenhouse gas emissions and diminished natural carbon sinks (IPCC, 2021).

2.2.2 Drivers of LULCC in Africa

Africa's urban population is increasing at a rate surpassing that of any other continent, with projections indicating urbanization rates will attain 60% by 2050 (UN-Habitat, 2022). Rapid urban growth significantly drives land use and land cover change (LULCC). Urban expansion frequently entails the transformation of forests, savannahs, and agricultural lands into residential and industrial areas.

Economic factors, including the demand for timber, fuelwood, and agricultural products, contribute to the exacerbation of forest loss in urban and peri-urban regions. Cities such as Nairobi and Lagos depend significantly on adjacent forests for fuelwood, which leads to deforestation and land degradation (African Forest Forum, 2022).

Inadequate policies contribute to land use and land cover change (LULCC). Inadequate land-use regulations and ineffective enforcement of conservation policies frequently facilitate uncontrolled urban expansion and unlawful logging activities. Informal settlements constitute a substantial segment of urban housing in Africa and are often located in ecologically sensitive regions as a result of insufficient urban planning (Wanjiru & Matsui, 2022).

Cultural factors, including traditional dependence on forests for livelihoods, significantly impact land use and land cover change (LULCC). These practices are sustainable in rural contexts; however, they become unsustainable in urban areas because of increased population densities and demand (FAO, 2023).

2.2.3 Sub-Saharan Africa: Drivers of urban LULCC

In Sub-Saharan Africa, urban land use and land cover change (LULCC) is primarily influenced by population growth and rural-to-urban migration. Cities such as Addis Ababa, Kampala, and Accra have experienced rapid expansion, resulting in the conversion of forests and agricultural lands into urban infrastructure (Myers et al., 2023). Economic activities, including mining, industrialization, and agriculture, significantly contribute to LULCC. In Ghana, for instance, gold mining in peri-urban areas has caused extensive deforestation, while in Ethiopia, urban agriculture has encroached upon forested lands (African Development Bank [AfDB], 2022). Political and governance challenges further exacerbate LULCC. Weak institutional frameworks, land tenure insecurity, and corruption hinder efforts to regulate land use and conserve natural ecosystems. Many urban forests in Sub-Saharan Africa lack formal protection, rendering them susceptible to encroachment (Global Forest Watch, 2022).

2.2.4 East Africa: Specific drivers of LULCC

The unique drivers of land use and land cover change in East Africa are influenced by its diverse landscapes and socio-economic dynamics. Urbanization in cities such as Nairobi, Kampala, and Dar es Salaam is driven by population growth and economic development. The expansion frequently takes place to the detriment of forests, wetlands, and agricultural lands (Wanjiru & Matsui, 2022).

Infrastructure development, including road construction and industrial parks, significantly influences land use and land cover change (LULCC) in East Africa. The Standard Gauge Railway in Kenya has led to considerable land conversion along its route, affecting adjacent ecosystems (Maathai, 2021).

Agricultural expansion in peri-urban regions constitutes a notable driver. The expansion of urban populations results in heightened food demand, prompting the transformation of forests into agricultural land. The situation in Tanzania illustrates this dynamic, as peri-urban forests such as Kazimzumbwi face encroachment due to small-scale farming activities (Mwakalila et al., 2023).

2.2.5 Tanzania: Drivers of urban LULCC

Urban land use and land cover change (LULCC) in Tanzania is influenced by demographic, economic, and policy factors. Rapid urbanization, especially in cities such as Dar es Salaam and Dodoma, has resulted in the transformation of natural forests into residential and commercial zones (MNRT, 2022). Informal settlements represent a major factor, as substantial populations reside in unplanned housing that intrudes upon forests and wetlands.

Agriculture and charcoal production significantly contribute to land use and land cover change (LULCC). Peri-urban forests frequently undergo clearance for agricultural purposes, and the significant demand for charcoal in urban regions contributes to deforestation in adjacent forested areas (Mgaya & Samwel, 2021).

The inadequate enforcement of land-use regulations intensifies land use and land cover change (LULCC). Tanzania possesses policies aimed at forest protection; however, resource limitations and corruption impede their effective implementation, resulting in ongoing illegal logging and encroachment (Nyika, 2020).

2.2.6 Kazimzumbwi Forest Reserve: Drivers of LULCC

The Kazimzumbwi Forest Reserve, located near Dar es Salaam, exemplifies the factors influencing land use and land cover change in Tanzania's peri-urban regions. Founded in 1936, the forest has experienced considerable degradation as a result of urbanization and human activities (Mwakalila et al., 2023).

S/n	Drivers of LULCC	Explanation
1.	Urbanization	The growth of Dar es Salaam has resulted in the intrusion of settlements into the forested areas. Informal housing and small-scale farming are common in the region, leading to deforestation and fragmentation of the forest landscape
2.	Economic activities	Charcoal production and illegal logging are major contributors to land use and land cover change in Kazimzumbwi. The significant demand for fuelwood in Dar es Salaam intensifies these activities, notwithstanding conservation efforts (Mgaya & Samwel, 2021).
3.	Challenges in policy and governance	The insufficient enforcement of conservation policies and the lack of community engagement have facilitated the continuation of illegal activities in Kazimzumbwi. Integration of forest ecosystems into urban planning frameworks is currently underway, yet it encounters considerable challenges (MNRT, 2022).

2.3 Impacts on Biodiversity and Ecosystem Functions

The degradation and loss of forests resulting from urbanization and land-use and land-cover change (LULCC) have substantial effects on biodiversity and ecosystem functions. Forests, especially in urban and peri-urban regions, serve as essential reservoirs of biodiversity and sources of ecosystem services. This section examines the impacts from a global viewpoint, subsequently focusing on Africa, Sub-Saharan Africa, East Africa, Tanzania, and the Kazimzumbwi Forest Reserve.

2.3.1 Global impacts on biodiversity and ecosystem functions

Forests globally host around 80% of terrestrial species, including animals, plants, and fungi, underscoring their importance for biodiversity (Food and Agriculture Organization [FAO], 2023). Urbanization and deforestation have resulted in habitat fragmentation, species loss, and disruptions to ecosystem processes. The global deforestation rate, estimated at 10 million hectares per year, has significantly reduced biodiversity in essential ecosystems, including the Amazon, Congo, and Southeast Asian rainforests (Hansen et al., 2021).

Urbanization impairs ecological connectivity, resulting in fragmented habitat patches that are less effective in sustaining wildlife populations. The fragmentation of habitats heightens the risk of extinction for species, especially those dependent on extensive habitats or possessing specialized ecological niches (Seto et al., 2012). Large mammals and migratory birds are significantly impacted by habitat fragmentation, resulting in cascading effects on ecosystems.

Forests contribute to ecosystem functions by offering services including carbon sequestration, water regulation, and soil stabilization. The reduction of forest cover diminishes these functions, intensifying global issues such as climate change, water scarcity, and soil degradation (IPCC, 2021).

2.3.2 Impacts on biodiversity and ecosystem functions in Africa

Africa contains several of the globe's most significant biodiversity hotspots, such as the Congo Basin and the coastal forests of East Africa. Urbanization and land use/land cover change are significantly degrading these ecosystems. From 2001 to 2021, Africa experienced a loss of around 36 million hectares of tree cover, which has considerably affected biodiversity and ecosystem functions (Global Forest Watch, 2022).

The loss of biodiversity in Africa is concerning, given the continent's reliance on natural ecosystems for livelihoods, agriculture, and cultural heritage. Urbanization has resulted in the displacement of essential species, diminished genetic diversity, and the degradation of ecosystem services, including pollination and water purification (African Forest Forum, 2022).

The replacement of forests with impervious surfaces compromises ecosystem functions. Urban heat islands in African cities are exacerbated by the reduction of tree cover, while deforestation diminishes forests' ability to regulate water cycles, resulting in heightened flooding and drought occurrences (UNEP, 2022).

2.3.3 Sub-Saharan Africa: Impacts on biodiversity and ecosystems

Sub-Saharan Africa (SSA) encounters distinct challenges in reconciling urban development with the preservation of biodiversity. The forests of the region host notable species including elephants, gorillas, and lions, which face escalating threats from habitat loss attributed to urban expansion (Myers et al., 2023). Urban sprawl in cities such as Kampala and Nairobi has intruded upon essential wildlife corridors, thereby disrupting animal movements and breeding patterns.

Deforestation and habitat degradation in Sub-Saharan Africa compromise essential ecosystem functions vital for human well-being. The forests in the region are essential for carbon sequestration, water supply maintenance, and the enhancement of soil fertility for agricultural purposes. The decline of these functions intensifies poverty and food insecurity, especially in urban and peri-urban regions (FAO, 2023).

2.3.4 East Africa: Regional impacts on biodiversity and ecosystems

East Africa is recognized for its biodiversity, featuring distinct ecosystems like the Eastern Arc Mountains and coastal forests. Urbanization and agricultural expansion are significantly degrading these landscapes. Cities such as Dar es Salaam, Nairobi, and Kampala have encroached upon forested regions, resulting in considerable biodiversity decline and diminished ecosystem functionality (Wanjiru & Matsui, 2022). Coastal forests are highly susceptible and offer essential services, including erosion control, habitat provision for marine species, and climate regulation. Their loss directly impacts fisheries, tourism, and climate resilience in the region (Maathai, 2021).

The fragmentation of forests in East Africa restricts their capacity to sustain biodiversity. Numerous endemic and threatened species, including the Udzungwa red colobus monkey in Tanzania, rely significantly on forest connectivity for their survival. The disruption of habitats results in population declines and, in certain instances, local extinctions (Nyika, 2020).

2.3.5 Tanzania: Biodiversity and ecosystem impacts

Tanzania's forests serve as biodiversity hotspots, hosting thousands of plant and animal species, a significant number of which are endemic. Urbanization, agricultural expansion, and illegal logging pose substantial threats to these ecosystems (Ministry of Natural Resources and Tourism [MNRT], 2022). From 2000 to 2020, Tanzania experienced a loss of about 20% of its forest cover, resulting in considerable effects on biodiversity and ecosystem functions (Global Forest Watch, 2022).

The decline in biodiversity in Tanzania is reflected in the decreasing populations of species including elephants, lions, and numerous bird species. Urban sprawl in cities such as Dar es Salaam and Dodoma leads to habitat destruction, which disrupts ecological networks and diminishes genetic diversity (Mgaya & Samwel, 2021).

The decline of ecosystem functions is a significant concern. Forests in Tanzania play a vital role in water catchment, especially in urban and peri-urban regions. Their degradation results in diminished water availability, increased soil erosion, and greater susceptibility to the impacts of climate change (Nyika, 2020).

2.3.6 Kazimzumbwi Forest Reserve: A case study

The Kazimzumbwi Forest Reserve, situated near Dar es Salaam, exemplifies the significant effects of urbanization on biodiversity and ecosystem functions. Founded in 1936, this coastal forest reserve serves as a crucial ecological resource for the region, hosting diverse flora and fauna, including endemic and threatened species (Mwakalila et al., 2023).

S/n	Impact	Explanation
1.	Biodiversity impacts	The expansion of urban areas and unlawful logging in Kazimzumbwi has resulted in habitat fragmentation and degradation. Species including the Zanzibar red colobus monkey and several bird species have undergone population declines as a result of habitat loss and degradation (Mgaya & Samwel, 2021).
2.	Ecosystem functions	Kazimzumbwi is essential for regulating the local climate, sequestering carbon, and recharging groundwater. Deforestation has diminished its ability to fulfill these functions, worsening urban issues like heat stress and water scarcity in Dar es Salaam (Nyika, 2020). The reduction of forest cover has led to heightened soil erosion, which impacts agricultural practices in adjacent regions.
3.	Conservation efforts	Efforts to restore and protect Kazimzumbwi are currently in progress, despite the existing challenges. Community-based conservation initiatives and reforestation programs seek to improve forest biodiversity and ecosystem functions. Incorporating Kazimzumbwi into the urban planning framework of Dar es Salaam may reduce the effects of land use and land cover change (LULCC) and improve urban resilience (MNRT, 2022).

PART II: ECOSYSTEM SERVICES OF URBAN FORESTS

CHAPTER THREE

3. BIOMASS AND CARBON SEQUESTRATION

3.1 Urban Forests as Carbon Sinks

Urban forests play a vital role in climate change mitigation by functioning as carbon sinks that absorb and store atmospheric carbon dioxide (CO_2). Forests significantly contribute to the reduction of urban greenhouse gas emissions and improve the resilience of cities against climate change. This section examines the role of urban forests as carbon sinks on a global scale, with an emphasis on Africa, Sub-Saharan Africa, East Africa, Tanzania, and specifically, the Kazimzumbwi Forest Reserve.

3.1.1 Global perspective: Urban forests as carbon sinks

Forests worldwide sequester around 2.6 billion metric tons of CO_2 each year, highlighting their critical role in global climate change mitigation efforts (FAO, 2023). Urban forests play a crucial role in carbon sequestration by absorbing CO_2 via photosynthesis and storing it in biomass and soil. Studies show that urban forests in cities such as New York, Tokyo, and London can sequester as much as 20% of urban CO_2 emissions, highlighting their critical role in reducing the urban carbon footprint (Nowak et al., 2021).

Urban forests contribute to indirect carbon savings through the reduction of energy consumption. Trees mitigate the demand for air conditioning by providing shade to buildings and cooling urban environments, which in turn decreases energy-related emissions. Strategically positioned trees in urban environments can decrease cooling energy consumption by as much as 30% (Nowak & Greenfield, 2022).

Urban forests face significant challenges due to rapid urbanization, deforestation, and climate change. The reduction of urban green spaces diminishes carbon sequestration capacity and intensifies climate-related challenges. Incorporating urban forests into climate action plans is crucial for meeting global carbon neutrality objectives (IPCC, 2021).

3.1.2 Urban forests as carbon sinks in Africa

In Africa, urban forests are gaining recognition for their role in carbon sequestration. The continent's urban areas are expanding rapidly, with urban populations projected to triple by 2050 (UN-Habitat, 2022). While this growth poses challenges, it also presents opportunities to enhance urban green infrastructure, including forests, to mitigate GHG emissions.

Urban forests in African cities like Nairobi, Accra, and Johannesburg sequester significant amounts of carbon, contributing to regional climate mitigation efforts. For instance, Nairobi's Karura Forest sequesters approximately 12,000 tons of CO_2 annually, highlighting the potential of urban forests as carbon sinks (Wanjiru & Matsui, 2022).

However, urbanization often leads to the loss of peri-urban forests, reducing their carbon storage capacity. Charcoal production, land conversion, and deforestation for construction exacerbate these losses. Addressing these challenges requires integrating urban forestry into city planning and promoting reforestation initiatives to restore degraded areas (African Forest Forum, 2022).

3.1.3 Sub-Saharan Africa: Carbon sequestration and urban forests

Sub-Saharan Africa (SSA) is undergoing considerable urban expansion, with urban areas encroaching upon natural ecosystems, such as forests. The expansion carries significant implications for carbon sequestration. Urban forests in Sub-Saharan Africa possess significant potential to mitigate urban emissions; however, they face growing threats from land-use changes and inadequate management (Global Forest Watch, 2022).

Studies indicate that peri-urban forests in Sub-Saharan Africa, including those in Addis Ababa, Ethiopia, and Kigali, Rwanda, sequester significant quantities of carbon in both biomass and soil. Forests serve as essential carbon sinks, contributing to the mitigation of emissions from transportation, industry, and energy sectors in urban environments (Myers et al., 2023).

Community-led reforestation initiatives are proving to be effective strategies for enhancing carbon sequestration in urban areas of Sub-Saharan Africa. Agroforestry practices in peri-urban regions of Uganda and Tanzania integrate tree planting with agricultural activities, enhancing carbon storage and food security (FAO, 2023).

3.1.4 East Africa: urban forests and carbon storage

Urban forests in East Africa, particularly in cities such as Kampala, Nairobi, and Dar es Salaam, are vital for regional carbon sequestration. These forests effectively absorb CO_2, mitigate heat stress, and deliver essential ecosystem services that enhance urban resilience (Nyika, 2020). Coastal forests, notably mangroves in Tanzania and Kenya, are especially significant for carbon sequestration, storing carbon at rates five times greater than terrestrial forests. This makes them crucial for climate change mitigation in urban coastal regions (Maathai, 2021). However, East Africa's urban forests face threats from deforestation, illegal logging, and land conversion. Protecting these forests necessitates the reinforcement of conservation policies, community engagement, and the integration of forests into urban planning frameworks (Wanjiru & Matsui, 2022).

3.1.5 Tanzania: Urban forests as carbon sinks

Urban forests in Tanzania play a crucial role in carbon sequestration and climate mitigation. Urban forests in Dar es Salaam, Dodoma, and Arusha play a crucial role in absorbing substantial quantities of CO_2, thereby mitigating emissions associated with rapid urbanization and industrialization (Ministry of Natural Resources and Tourism [MNRT], 2022).

Peri-urban forests, including the Pugu and Kazimzumbwi Forest Reserves, serve as essential carbon sinks for urban regions in Tanzania. Forests sequester carbon in both biomass and soil, while also offering benefits including biodiversity conservation and climate regulation (Mgaya & Samwel, 2021).

Urban forests in Tanzania face challenges such as deforestation for fuelwood, agricultural expansion, and the proliferation of unplanned urban settlements. To address these challenges, coordinated efforts among government agencies, non-governmental organizations, and local communities are essential for promoting sustainable forest management and reforestation (Nyika, 2020).

3.1.6 Kazimzumbwi Forest Reserve: A case study

Adopting LULCC in Table 2.2 for the years 1994, 2004, and 2024 for Kazimzumbwi Forest Reserve (KFR) and Figure 3.1 below shows the flow chart of the methodological approach used for the estimation of the biomass and carbon stocks for aforementioned years and the computation of changes between studies periods.

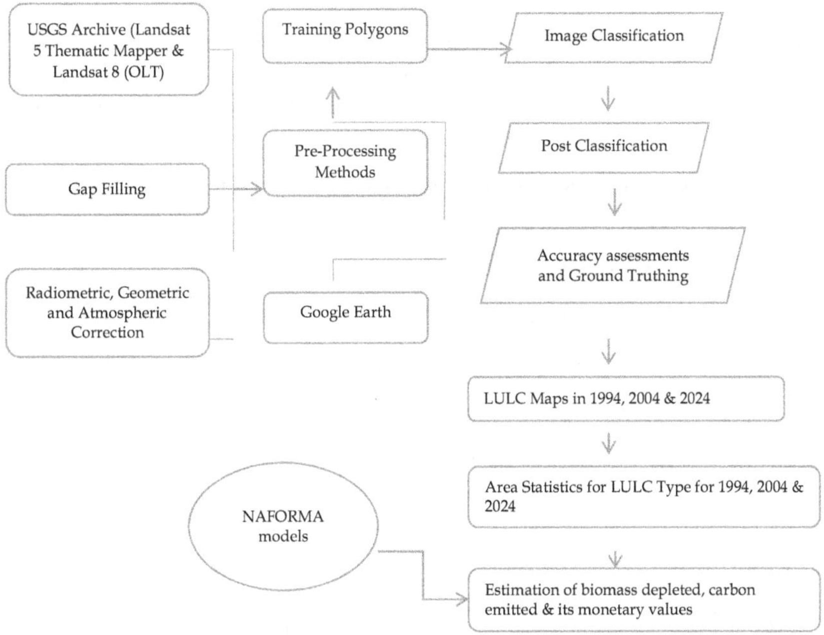

Figure 3.1: Flowchart of the methodological approach for this study

Data analysis

Amount of biomass depleted as a result of LULCC in KFR from 1994 to 2024

Tanzania forest Carbon can be estimated in three pools namely AGB (above ground biomass), BGB (below ground biomass) and DW (dead wood) (URT, 2015). BGB was estimated as a fraction of AGB. AGB and BGB were estimated as follows:

(i) AGB (tonnes/ha) = Tree stem volume (m^3/ha) * wood density/1000; and

(ii) BGB (tonnes/ha) = AGB * 0.25 (as default), or root to shoot ratios.

URT (2015) uses conversion factors into programmed NAFORMA analysis system by tree species or species groups to provide standards in each terrestrial ecosystem of Tanzania. Dead wood (DW) biomass is estimated from the volume computed using

Smalian formula multiplied by wood density of 619 kg/m^3 (Chidumayo, 2012 cited by URT, 2015). URT (2015) through NAFORMA reveals the dead wood Biomass of Tanzania (Table 3.2) is relatively low since deadest wood in accessible areas is collected as fuelwood. As woodlands are generally more accessible than forests, collection of deadwood for fuelwood from these areas is easier as shown in Table 3.1.

Table 3.1: Living tree stemwood and deadwood biomass by primary vegetation type

LULC type	Change detected (ha) 1994-2024	Biomass (t/ha) AGB	BGB	DWB	Total biomass
Closed forest	1835.64	59.5	18.2	5.09	82.79
Open woodland	9.56	27.7	9.5	1.89	39.09
Shrubs	-788.7	11	4.4	0.73	16.13
Grassland	-689.33	2.9	1.1	0.36	4.36
Bare land	-294.94	2.9	1.1	0.36	4.36
Built-up	-72.23	2.9	1.1	0.36	4.36
Total		**106.9**	**35.4**	**8.79**	**151.09**

Amount of carbon emitted to the atmosphere as a result of LULCC in KFR from 1994 to 2024

According to URT (2015), carbon in terrestrial ecosystems of Tanzania can be computed as follows:

Carbon (tonnes/ha) = Biomass * 0.47

Living tree stemwood (AGC + BGC) and dead wood carbon (DWC) (t/ha) by primary vegetation type are illustrated in Table 3.2.

Table 3.2: Living tree stemwood and deadwood carbon by primary vegetation type

LULC type	Carbon (t/ha)			
	AGC	BGC	DWC	Total carbon
Closed forest	27.97	8.55	2.39	38.91
Open woodland	13.02	4.47	0.89	18.37
Shrubs	5.17	2.07	0.34	7.58
Grassland	1.36	0.52	0.17	2.05
Bare land	1.36	0.52	0.17	2.05
Built-up	1.36	0.52	0.17	2.05
Total	**50.24**	**16.64**	**4.13**	**71.01**

3.1.7 Carbon dioxide emitted to the atmosphere

Estimating the carbon dioxide (CO_2) emitted from Kazimzumbwi Forest Reserve (KFR) involves converting total carbon emitted into CO_2 equivalents, crucial for understanding its climate mitigation role. The Intergovernmental Panel on Climate Change (IPCC) provides guidelines for this conversion, using a factor of 3.67, representing the molecular weight ratio of CO_2 to carbon (IPCC, 2006 & 2023). The formula is:

$$CO_2 \text{ (tonnes)} = \text{Carbon (tonnes)} \times 3.67$$

Applying this to the estimated carbon stocks of living and dead biomass in KFR gives the CO_2 sequestration potential as shown I Table 3.3:

$$\text{Total } CO_2 \text{ (tonnes)} = AGCO_2 \text{ (tonnes)} + BGCO_2 \text{ (tonnes)} + DWCO_2 \text{ (tonnes)}$$

Table 3.3: Living tree stemwood and deadwood carbon dioxide by primary vegetation type

LULC type	Carbon dioxide (t/ha)			
	$AGCO_2$	$BGCO_2$	$DWCO_2$	Total CO_2
Closed forest	102.63	31.39	8.78	142.80
Open woodland	47.78	16.39	3.26	67.43
Shrubs	18.97	7.59	1.26	27.82
Grassland	5.00	1.90	0.62	7.52
Bare land	5.00	1.90	0.62	7.52
Built-up	5.00	1.90	0.62	7.52
Total	**184.39**	**61.06**	**15.16**	**260.62**

Economic loss of KFR resulted from carbon emissions from 1994 to 2024

The study adopted from Jenkins (2014), and Lobora *et al.* (2017) emphasized that, the standard carbon market is US$ 4 per ton if REDD+ is implemented; this was used to estimate financial cost of carbon emitted to the atmosphere as a result of land use/cover change in KFR for the period 1994-2024 as indicated in Table 3.4.

Table 3.4: Economic loss/gain of KFR from carbon dioxide trade by primary vegetation type

LULC type	Carbon dioxide (US$/ha)			
	$AGCO_2$	$BGCO_2$	$DWCO_2$	Total CO_2
Closed forest	410.53	125.57	35.12	571.22
Open woodland	191.12	65.55	13.04	269.71
Shrubs	75.90	30.36	5.04	111.29
Grassland	20.01	7.59	2.48	30.08
Bare land	20.01	7.59	2.48	30.08
Built-up	20.01	7.59	2.48	30.08
Total	**737.57**	**244.25**	**60.65**	**1042.46**

Results and Discussion

Biomass depleted as a result of LULCC in KFR from 1994 to 2024

Urban forests play a crucial role in carbon sequestration, biodiversity conservation, and providing essential ecosystem services that contribute to urban resilience and human well-being (Zhao et al., 2016; McDonald et al., 2020). However, the Kazimzumbwi Forest Reserve (KFR) has experienced significant biomass loss over three decades, with reductions of 151,972.64 tons in closed forests and 373.70 tons in open woodlands, resulting in a net loss of 135,018.27 tons from 1994 to 2024. In contrast, non-forest land uses such as shrubs and grasslands have seen biomass increases, reflecting ongoing urban expansion and agricultural encroachment (Seto et al., 2012).

The degradation of KFR highlights the alarming decline of critical ecosystems that provide carbon storage and biodiversity (Pan et al., 2011; IPCC, 2021). The shift from high-biomass ecosystems to lower-biomass land uses diminishes ecological integrity and resilience, exacerbating climate change impacts and reducing habitat stability (Pimm et al., 2014). Socio-economically, the depletion of forest resources threatens livelihoods dependent on forestry products and services, particularly for marginalized communities (FAO, 2018). Additionally, urbanization intensifies the urban heat island effect and deteriorates air quality, adversely affecting public health (Loughner et al., 2012). To combat these challenges, integrated urban forest management strategies are essential. Restoration efforts, such as afforestation and reforestation, can recover lost biomass and enhance ecosystem services (Chazdon et al., 2020). Sustainable land-use planning and community engagement in conservation efforts are also critical for fostering stewardship and resilience (Adhikari et al., 2014). Overall, protecting and restoring urban forests like KFR is vital for maintaining ecological integrity and supporting the socio-economic well-being of urban populations, necessitating collaboration among policymakers, urban planners, and stakeholders.

Table 3.5: KFR's living tree stemwood and deadwood biomass depleted from 1994 to 2024

LULC type	Biomass (t)			
	AGB	BGB	DWB	Total biomass
Closed forest	109,220.58	33,408.65	9,343.41	151,972.64
Open woodland	264.81	90.82	18.07	373.70
Shrubs	(8,675.70)	(3,470.28)	(575.75)	(12,721.73)
Grassland	(1,999.06)	(758.26)	(248.16)	(3,005.48)

Bare land	(855.33)	(324.43)	(106.18)	(1,285.94)
Built-up	(209.47)	(79.45)	(26.00)	(314.92)
Total	**97,745.84**	**28,867.04**	**8,405.39**	**135,018.27**

( ) = -ve

Carbon emitted to the atmosphere as a result of LULCC in KFR from 1994 to 2024

Urban forests are essential for ecological balance and socio-economic benefits in urban areas, acting as carbon sinks, mitigating heat, supporting biodiversity, and enhancing urban life quality (McPherson et al., 2017; Nowak et al., 2020). However, rapid urbanization and unsustainable land-use practices have led to significant degradation of these ecosystems. This analysis focuses on carbon dynamics in the Kazimzumbwi Forest Reserve (KFR) from 1994 to 2024, revealing critical implications for ecological integrity and socio-economic sustainability. The study findings indicate substantial changes in carbon stocks across various land-use categories within KFR over the 30-year period. Closed forests experienced a carbon loss of 71,427.14 tons, while open woodlands saw a minor reduction of 175.64 tons. In contrast, carbon gains were noted in shrubs, grasslands, bare land, and built-up areas, totaling 5,979.21 tons, 1,412.58 tons, 604.39 tons, and 148.01 tons, respectively. Despite these gains, the net carbon loss for KFR during the study period amounted to 63,458.58 tons.

These results highlight the vulnerability of urban forests to human pressures, emphasizing the ecological consequences of forest degradation and land-use change. The significant carbon loss in closed forests is attributed to deforestation, forest fragmentation, urban expansion, agricultural encroachment, and unsustainable resource extraction (Seto et al., 2012; IPCC, 2021). Although non-forest categories like shrubs and grasslands showed carbon gains, their lower carbon storage capacities compared to mature forests render these gains insufficient to offset the losses from high-carbon ecosystems (Chapin et al., 2000). The net carbon loss of 63,458.58 tons over three decades signifies a substantial reduction in KFR's carbon sequestration capacity, undermining urban forests' role in climate change mitigation (Pan et al., 2011). Additionally, forest degradation leads to habitat loss and reduced biodiversity, further weakening urban ecosystem resilience (Pimm et al., 2014).

Socio-economically, the degradation of urban forests like KFR affects local livelihoods, particularly among vulnerable communities reliant on forest resources such as timber and fuelwood (FAO, 2018). The decline in forest cover exacerbates environmental issues like urban heat islands and air pollution, negatively impacting public health and urban livability (Loughner et al., 2012). Furthermore, the loss of ecosystem services diminishes the cultural and recreational value of urban forests, essential for urban populations' well-being (McDonald et al., 2020).

To mitigate ongoing carbon losses in KFR, sustainable forest management practices and restoration initiatives are crucial. Strategies include reforestation and afforestation, urban land-use planning, community engagement, and policy implementation to prevent illegal activities contributing to forest degradation. Prioritizing urban forest conservation and restoration is vital for sustaining ecological and socio-economic functions amidst urbanization and climate change. Collaborative efforts among policymakers, urban planners, and stakeholders are essential for developing and implementing effective strategies (Chazdon et al., 2020; Adhikari et al., 2014).

Table 3.6: Living tree stemwood and dead wood carbon emitted from 1994 to 2024

LULC type	Carbon (t)			
	AGC	BGC	DWC	Total carbon
Closed forest	51,333.67	15,702.06	4,391.40	71,427.14
Open woodland	124.46	42.69	8.49	175.64
Shrubs	(4,077.58)	(1,631.03)	(270.60)	(5,979.21)
Grassland	(939.56)	(356.38)	(116.63)	(1,412.58)
Bare land	(402.00)	(152.48)	(49.90)	(604.39)
Built-up	(98.45)	(37.34)	(12.22)	(148.01)
Total	**45,940.55**	**13,567.51**	**3,950.53**	**63,458.58**

Carbon dioxide emitted to the atmosphere as a result of LULCC in KFR from 1994 to 2024

Urban forests are crucial in combating climate change by acting as carbon sinks, absorbing atmospheric carbon dioxide (CO_2) through photosynthesis and storing it as biomass (Pan et al., 2011; Nowak et al., 2020). However, degradation or conversion of these forests leads to the release of stored carbon, exacerbating greenhouse gas emissions. This analysis focuses on the CO_2 emissions from biomass changes in the Kazimzumbwi Forest Reserve (KFR) from 1994 to 2024, revealing significant emissions from closed forests (262,137.60 tons) and open woodlands (644.60 tons), while other land types sequestered a total of 29,889.19 tons, resulting in net emissions of 232,893.01 tons.

The findings indicate that forest degradation significantly impacts carbon dynamics in urban forests. Closed forests, with their dense biomass, are highly effective at sequestering carbon, and their loss not only releases CO_2 but also diminishes future carbon storage capacity (IPCC, 2021). The lower emissions from open woodlands reflect their reduced biomass density, yet their degradation still contributes to overall carbon loss (Chapin et al., 2000). Gains in shrubs and grasslands suggest some recovery of vegetation, but their carbon sequestration potential is much lower than that of mature forests, and urban vegetation faces additional stressors that further limit its effectiveness (McDonald et al., 2020).

Ecologically, the net emissions from KFR have serious implications for urban ecosystems and global climate efforts, as urban forests play a vital role in the carbon cycle and their degradation increases atmospheric CO_2 levels (Seto et al., 2012). The loss of forest cover also threatens biodiversity and disrupts essential ecosystem services, such as water regulation and temperature control (Pimm et al., 2014).

Socio-economically, urban forest degradation impacts livelihoods, particularly for marginalized communities reliant on forest resources (FAO, 2018; Adger et al., 2003). The conversion of forests to urban areas exacerbates the urban heat island effect and diminishes air quality, affecting public health and access to recreational spaces (Loughner et al., 2012; McPherson et al., 2017).

To mitigate these emissions, strategies such as reforestation, sustainable land-use planning, carbon offset programs, community engagement, and policy enforcement are essential (Chazdon et al., 2020; Seto et al., 2012; Adhikari et al., 2014). These measures can enhance carbon sequestration and preserve the ecological and socio-economic integrity of urban forests amidst increasing urbanization and climate change pressures.

Table 3.7: Carbon dioxide emitted to the atmosphere from 1994 to 2024

LULC type	Carbon dioxide (t)			
	AGCO$_2$	BGCO$_2$	DWCO$_2$	Total CO$_2$
Closed forest	188,394.58	57,626.58	16,116.44	262,137.60
Open woodland	456.77	156.66	31.17	644.60
Shrubs	(14,964.71)	(5,985.89)	(993.11)	(21,943.71)
Grassland	(3,448.17)	(1,307.93)	(428.05)	(5,184.15)
Bare land	(1,475.35)	(559.62)	(183.15)	(2,218.12)
Built-up	(361.31)	(137.05)	(44.85)	(543.21)
Total	**168,601.80**	**49,792.75**	**14,498.45**	**232,893.01**

Economic loss of KFR resulted from carbon emissions from 1994 to 2024

The economic analysis of the Kazimzumbwi Forest Reserve (KFR) from 1994 to 2024 reveals significant financial implications tied to changes in carbon stocks, highlighting the economic losses associated with forest degradation and land-use changes. The findings indicate a substantial economic loss of US$ 1,048,550.40 from closed forests, which are high-biomass ecosystems that contribute significantly to carbon emissions. Open woodlands added a smaller loss of US$ 2,578.38 due to their lower biomass density. Conversely, areas experiencing carbon stock increases, such as shrubs, grasslands, bare lands, and built-up areas, generated profits totaling US$ 119,556.80. However, the overall net economic loss for KFR in the carbon market over the 30-year period amounted to US$ 931,572.02 (Stern, 2006; IPCC, 2021).

The analysis emphasizes the economic consequences of forest degradation, particularly in high-biomass areas. The loss of over US$ 1 million from closed forests underscores the direct economic impact of deforestation and the missed opportunities for carbon credits through conservation efforts. While the profits from secondary vegetation indicate potential economic benefits, they are insufficient to offset the losses from mature forests (Chapin et al., 2000; Pan et al., 2011).

The implications for urban carbon economics are profound, as the degradation of urban forests undermines their financial value and their role in mitigating greenhouse gas emissions (McPherson et al., 2017; Nowak et al., 2020). The loss of potential revenue from carbon credits also affects funding opportunities for conservation initiatives, suggesting that forests like KFR could serve as financial resources for local communities through carbon trading schemes (FAO, 2018).

Beyond carbon market implications, the socio-economic value of urban forests encompasses ecosystem services such as air purification and recreational opportunities, which are crucial for low-income communities that rely on these resources for livelihoods and resilience (Adger et al., 2003). The transition of forested land to other uses often prioritizes short-term economic gains over sustainability, further diminishing the ecological and social benefits of forests.

To address these challenges, recommendations include engaging KFR in carbon trading schemes, investing in reforestation, integrating carbon market considerations into urban planning, involving local communities in forest management, and promoting the multi-functional value of urban forests. By leveraging carbon trading and addressing forest degradation, it is possible to enhance the ecological and socio-economic integrity of urban forests, supporting climate change mitigation and long-term sustainability.

Table 3.8: Economic loss of KVFP resulted from carbon emissions from 1994 to 2024

LULC type	Carbon dioxide (US$)			
	AGCO$_2$	BGCO$_2$	DWCO$_2$	Total CO$_2$
Closed forest	753,578.31	230,506.31	64,465.78	1,048,550.40
Open woodland	1,827.10	626.62	124.66	2,578.38
Shrubs	(59,858.86)	(23,943.54)	(3,972.45)	(87,774.86)
Grassland	(13,792.69)	(5,231.71)	(1,712.20)	(20,736.60)
Bare land	(5,901.41)	(2,238.46)	(732.59)	(8,872.46)
Built-up	(1,445.24)	(548.19)	(179.41)	(2,172.84)
Total	**674,407.21**	**199,171.02**	**57,993.79**	**931,572.02**

3.2 Measuring and Monitoring Biomass

Biomass, defined as the total mass of living organisms within a specified area, serves as a vital indicator of forest health, carbon sequestration capacity, and ecosystem functionality. Precise measurement and monitoring of biomass are crucial for comprehending forest dynamics, evaluating their function as carbon sinks, and executing effective conservation strategies. This section explores the methodologies and challenges of measuring and monitoring biomass, from global perspectives to specific cases in Africa, Sub-Saharan Africa, East Africa, Tanzania, and the Kazimzumbwi Forest Reserve.

3.2.1 Global perspective on measuring and monitoring biomass

Measuring biomass is essential for effective forest management and climate mitigation strategies worldwide. Quantifying biomass offers insights into carbon storage capacity, forest productivity, and ecosystem services (Food and Agriculture Organization [FAO], 2023). Biomass measurement methods can be classified into two main categories: direct and indirect approaches.

Direct methods entail the collection and weighing of vegetation, yielding accurate measurements; however, they are labor-intensive and can be destructive (Chave et al., 2014). Such studies are frequently constrained to small scales.

Indirect methods utilize non-destructive techniques, which include:

(i) Allometric equations utilize tree dimensions, including diameter at breast height (DBH) and tree height, to estimate biomass (Brown, 1997).

(ii) Remote sensing employs satellite imagery, LiDAR (Light Detection and Ranging), and UAV (Unmanned Aerial Vehicle) technologies for the extensive monitoring of biomass (Saatchi et al., 2021).

(iii) Monitoring biomass is essential for global initiatives such as REDD+ (Reducing Emissions from Deforestation and Forest Degradation), which focus on quantifying carbon stocks and mitigating deforestation (UNFCCC, 2022).

3.2.2 Measuring and monitoring biomass in Africa

The diverse forest ecosystems of Africa are integral to both regional and global carbon cycles. The continent encounters considerable challenges in the measurement and monitoring of biomass, attributed to constrained resources and technical capacity (African Forest Forum, 2022). Tropical forests in the Congo Basin store around 60 billion tons of carbon; however, monitoring efforts are limited due to inadequate infrastructure (Global Forest Watch, 2022).

Remote sensing technologies are progressively utilized for monitoring biomass in African forests. Satellite data from programs such as NASA's Landsat and the European Space Agency's Sentinel satellites have enhanced biomass estimation throughout the continent (Saatchi et al., 2021). Ground-based measurements are crucial for the validation of remote sensing data and the development of region-specific allometric equations.

Community-based monitoring initiatives are increasingly being adopted in Africa. This entails educating local communities to gather data on tree measurements and forest conditions, addressing deficiencies in technical knowledge and fostering participatory conservation (Wanjiru & Matsui, 2022).

3.2.3 Biomass monitoring in Sub-Saharan Africa

Sub-Saharan Africa (SSA) encompasses diverse forest types, such as tropical rainforests, miombo woodlands, and coastal forests. Each ecosystem necessitates specific methodologies for biomass assessment. For instance:

(i) Biomass monitoring in Miombo Woodlands employs allometric equations tailored to the predominant tree species, including Brachystegia and Julbernardia (Chidumayo et al., 2020).

(ii) Remote sensing and LiDAR technologies have proven effective in quantifying biomass in dense tropical forests, including those in the Congo Basin (Saatchi et al., 2021).

(iii) Coastal forests depend on ground-based and drone technologies for biomass monitoring due to their fragmented characteristics (Myers et al., 2023).

Despite advancements, Sub-Saharan Africa encounters challenges including insufficient funding, restricted technical expertise, and an absence of standardized methodologies for biomass monitoring. Addressing these issues is essential for enhancing data accuracy and facilitating forest conservation initiatives.

3.2.4 East Africa: Advances in biomass monitoring

East Africa has achieved notable progress in the adoption of contemporary technologies for biomass measurement. Countries such as Kenya, Uganda, and Tanzania utilize satellite imagery and Geographic Information Systems (GIS) to assess forest biomass (Nyika, 2020). These tools are essential for tracking biomass changes over time and for pinpointing regions of deforestation and degradation.

Community participation is a fundamental aspect of biomass monitoring in East Africa. Initiatives such as the Green Belt Movement in Kenya engage local communities in data collection, thereby improving both accuracy and conservation results (Maathai, 2021). Challenges, including data gaps and inconsistent methodologies, continue to exist.

Mangrove forests, integral to East Africa's coastal ecosystems, are significant for biomass monitoring. These forests sequester considerably greater amounts of carbon per hectare compared to terrestrial forests, thus rendering their monitoring essential for regional climate strategies (Wanjiru & Matsui, 2022).

3.2.5 Biomass monitoring in Tanzania

Biomass monitoring in Tanzania is essential for comprehending the ecological and socio-economic functions of forests. The forests of the country, encompassing miombo woodlands and coastal forests, are estimated to sequester around 1.5 billion tons of carbon (Ministry of Natural Resources and Tourism [MNRT], 2022).

Tanzania utilizes remote sensing, ground-based measurements, and community-based methods for biomass monitoring. For example:

(i) Remote sensing employs satellite data to delineate forest cover and assess biomass variations across extensive regions (Global Forest Watch, 2022).

(ii) Field surveys quantify tree dimensions to calibrate and validate remote sensing data.

(iii) Local communities are essential in the collection of biomass data, especially in rural and peri-urban regions (Nyika, 2020).

Tanzania continues to encounter challenges, including deforestation, illegal logging, and restricted technical capacity. Enhancing biomass monitoring systems is crucial for the advancement of conservation policies and adherence to international commitments such as REDD+ (MNRT, 2022).

3.2.6 Kazimzumbwi Forest Reserve: biomass monitoring

The Kazimzumbwi Forest Reserve, situated near Dar es Salaam, illustrates the significance of biomass monitoring in peri-urban forest ecosystems. Kazimzumbwi, one of Tanzania's oldest coastal forests, is essential for carbon sequestration and biodiversity conservation (Mwakalila et al., 2023).

S/n	Biomass monitoring	Explanation
1.	Methods of biomass monitoring	Biomass monitoring in Kazimzumbwi integrates remote sensing techniques with ground-based measurements. Satellite imagery from programs such as Landsat and Sentinel is utilized to evaluate changes in forest cover, whereas field surveys yield data on tree species, dimensions, and density (Mgaya & Samwel, 2021).
2.	Challenges	Encroachment, illegal logging, and insufficient funding impede effective biomass monitoring in Kazimzumbwi. Such activities degrade the forest and diminish its carbon storage capacity and ecosystem functions (Nyika, 2020).
3.	Opportunities	Community-based monitoring initiatives and collaborations with research institutions provide avenues for improving biomass measurement in Kazimzumbwi. The integration of modern technologies, including drones and LiDAR, has the potential to enhance data accuracy and bolster conservation initiatives (MNRT, 2022).

CHAPTER FOUR

4. CLIMATE REGULATION AND CO$_2$ REDUCTION

4.1 Role in Mitigating Urban Heat Islands

Urban Heat Islands (UHIs) are urban regions that exhibit markedly elevated temperatures compared to adjacent rural areas, attributable to anthropogenic activities, increased impervious surfaces, and diminished vegetation. Urban forests are essential in mitigating urban heat islands by offering shade, facilitating cooling via evapotranspiration, and improving air quality. This section analyzes the role of forests in mitigating urban heat islands (UHIs), assessing their effects from a global perspective to the specific instance of Kazimzumbwi Forest Reserve in Tanzania.

4.1.1 Global role of forests in mitigating UHIs

Urban forests are acknowledged worldwide as essential green infrastructure for mitigating the negative impacts of urban heat islands (UHIs). Trees and green spaces mitigate surface and air temperatures by providing shade to impervious surfaces and facilitating water vapor release via evapotranspiration. Research indicates that urban forests can reduce temperatures by 2–4°C in densely populated urban environments, providing considerable alleviation from heat stress (Nowak et al., 2021).

Urban forests decrease energy consumption by cooling buildings, thereby reducing the need for air conditioning and the related greenhouse gas emissions. Cities such as Singapore and New York have incorporated urban forestry into their climate action strategies, illustrating the effectiveness of tree planting and preservation in mitigating urban heat islands (Tan et al., 2022).

The effectiveness of urban forests in mitigating urban heat islands is contingent upon tree species, canopy density, and spatial distribution. Cities worldwide are implementing strategies to enhance the cooling benefits of urban forests, including increasing canopy cover, restoring degraded green spaces, and establishing green corridors (FAO, 2023).

4.1.2 Mitigating UHIs in Africa

Africa's swift urbanization is exacerbating the urban heat island effect, especially in densely populated cities with inadequate green infrastructure. Urban forests in Africa are essential for mitigating urban heat islands, providing cooling effects, improving air quality, and strengthening urban resilience. Research conducted in Nairobi, Kenya, indicates that urban green spaces lower temperatures in adjacent neighborhoods by an average of 3°C (Wanjiru & Matsui, 2022).

Urban forests play a crucial role in mitigating urban heat islands (UHIs) in African cities, particularly as heatwaves increase in frequency due to climate change. Many urban forests in Africa face threats from urban sprawl, deforestation, and insufficient urban planning, despite their advantages. The African Green Cities Initiative seeks to improve green infrastructure and incorporate forests into urban planning to tackle these issues (UNEP, 2022).

4.1.3 Sub-Saharan Africa: Addressing UHIs with urban forests

Sub-Saharan Africa (SSA) is witnessing significant urbanization rates, which are contributing to the growth of urban heat islands (UHIs). Urban forests in Sub-Saharan Africa play a critical role in mitigating these effects by offering natural cooling and decreasing the thermal load in urban environments. Addis Ababa, Ethiopia, has implemented urban forestry programs to address heat stress, illustrating the advantages of tree planting and preservation for urban environments (Myers et al., 2023).

Challenges in Sub-Saharan Africa encompass inadequate resources for the upkeep of urban forests, insufficient policy frameworks, and the encroachment upon green spaces. Community-based conservation and reforestation initiatives are proving to be effective strategies for improving the role of urban forests in mitigating urban heat islands. These programs frequently engage local communities in tree planting and maintenance, thereby ensuring the sustainability of green spaces (African Development Bank [AfDB], 2022).

4.1.4 East Africa: Urban forests and UHI mitigation

Urban heat islands (UHIs) are increasingly impacting East Africa's urban centers, such as Nairobi, Kampala, and Dar es Salaam, as a result of rapid urbanization and deforestation. Urban forests in the region play a vital role in mitigating temperatures and enhancing urban living conditions. Research conducted in Nairobi's Karura Forest indicates that regions with dense tree cover exhibit markedly lower temperatures compared to adjacent urban areas (Wanjiru & Matsui, 2022).

In coastal cities such as Dar es Salaam, forests contribute to the mitigation of urban heat islands and humidity effects. Mangroves and coastal forests offer cooling benefits and serve as protective barriers for urban regions against sea-level rise and storm surges (Maathai, 2021). The loss of green spaces resulting from land conversion and unregulated urban expansion constitutes a significant challenge.

4.1.5 Tanzania: Role of Urban Forests in UHI Mitigation

The urban forests of Tanzania, especially in Dar es Salaam, Dodoma, and Arusha, play a crucial role in mitigating urban heat islands (UHIs). Dar es Salaam, the largest and fastest-growing city in Tanzania, faces significant heat stress, intensified by high population density and insufficient green infrastructure. Urban forests and green spaces in cities offer critical cooling services, lowering surface temperatures and improving urban resilience (Ministry of Natural Resources and Tourism [MNRT], 2022).

The Pugu and Kazimzumbwi Forest Reserves function as ecological buffers, contributing to the cooling of adjacent areas and enhancing air quality. These forests face threats from encroachment, illegal logging, and insufficient conservation measures. Enhancing policies for the protection of urban forests and their integration into urban planning is essential for addressing urban heat islands in Tanzania (Nyika, 2020).

4.1.6 Kazimzumbwi Forest Reserve: A case study

The Kazimzumbwi Forest Reserve, situated near Dar es Salaam, serves as a crucial natural resource for alleviating urban heat islands in Tanzania's largest city. Kazimzumbwi, one of the oldest coastal forests in the region, offers substantial ecological and socio-economic advantages, such as temperature regulation and climate resilience (Mwakalila et al., 2023).

S/n	UHIs mitigation	Explanation
1.	Cooling effects	The dense canopy and biodiversity of Kazimzumbwi enhance its capacity to reduce surrounding temperatures. The forest functions as a natural air conditioning system, reducing the urban heat island effect in adjacent urban regions. The cooling benefits are applicable to peri-urban settlements, improving residents'

		quality of life and mitigating heat stress during extreme weather events (Mgaya & Samwel, 2021).
2.	Challenges and threats	Kazimzumbwi encounters substantial challenges, notably deforestation, encroachment, and insufficient funding for conservation efforts. The threats diminish the forest's capacity to deliver cooling effects and other ecosystem services, thereby intensifying UHI impacts in Dar es Salaam (MNRT, 2022).
3.	Conservation opportunities	Community-based reforestation initiatives and collaborations with research institutions present opportunities for the restoration of Kazimzumbwi's ecological integrity. Incorporating the forest into the urban planning framework of Dar es Salaam may improve its capacity to mitigate urban heat islands and promote sustainable urban development (Nyika, 2020).

4.2 Carbon Dioxide Absorption and Air Quality Improvement

Urban forests function as essential green infrastructure, playing a significant role in reducing carbon dioxide (CO_2) concentrations and enhancing air quality. Urban forests utilize photosynthesis to absorb CO_2, a major greenhouse gas that contributes to climate change. They also filter air pollutants such as particulate matter (PM), nitrogen dioxide (NO_2), and ozone (O_3), thereby improving urban air quality and mitigating health risks. This section examines the function of forests in carbon dioxide absorption and the enhancement of air quality, beginning with a global overview and progressively focusing on Africa, Sub-Saharan Africa, East Africa, Tanzania, and ultimately, the Kazimzumbwi Forest Reserve.

4.2.1 Global Role of forests in CO_2 absorption and air quality

Forests globally serve as essential carbon sinks, sequestering around 2.6 billion tons of CO_2 each year (Food and Agriculture Organization [FAO], 2023). Urban forests play a crucial role in sequestering carbon within biomass and soils. Studies indicate that an individual mature tree can absorb approximately 22 kilograms of CO_2 each year, whereas a hectare of forest has the capacity to sequester between 1 to 5 tons of CO_2 annually, contingent upon tree species and environmental factors (Nowak et al., 2021).

Urban forests enhance air quality through the filtration of pollutants. Tree canopies capture airborne particulates, including PM2.5 and PM10, and absorb gaseous pollutants such as NO_2, O_3, and sulfur dioxide (SO_2). Studies in New York City estimate that urban trees remove approximately 2,000 tons of air pollutants annually, which contributes to a reduction in respiratory illnesses and an improvement in public health (Nowak et al., 2022).

Rapid urbanization and deforestation, however, jeopardize the capacity of forests to fulfill these functions. Global initiatives, including the United Nations' Sustainable Development Goals (SDG 11: Sustainable Cities and Communities), highlight the necessity of incorporating green infrastructure into urban planning to improve air quality and decrease carbon emissions (United Nations, 2022).

4.2.2 CO_2 absorption and air quality improvement in Africa

Africa's forests serve as crucial global carbon sinks, with the Congo Basin sequestering more than 600 million tons of CO_2 each year (Global Forest Watch, 2022). Urban forests in African cities are essential for regulating local and regional climates and for purifying air. Forests in urban areas such as Nairobi, Accra, and Johannesburg significantly absorb CO_2, thereby mitigating emissions from transportation, industry, and residential energy consumption (Wanjiru & Matsui, 2022).

Urban forests in Africa enhance air quality by reducing pollution from expanding urban centers. Urban trees in Lagos, Nigeria, have demonstrated a capacity to reduce particulate pollution by as much as 15%, thereby mitigating health risks linked to respiratory and cardiovascular diseases (African Development Bank [AfDB], 2022). Challenges including deforestation, land degradation, and insufficient urban planning hinder the capacity of forests to absorb CO_2 and enhance air quality.

4.2.3 Sub-Saharan Africa: Carbon sequestration and air quality

Sub-Saharan Africa (SSA) is experiencing escalating air pollution issues attributed to urbanization, industrialization, and dependence on biomass for energy. Urban forests in Sub-Saharan Africa play a crucial role in carbon dioxide absorption and pollutant filtration. Peri-urban forests in Addis Ababa and Kigali enhance local air quality by removing pollutants and offering cooling effects (Myers et al., 2023).

The ability of urban forests in SSA to sequester carbon and enhance air quality is frequently compromised by deforestation, illegal logging, and encroachment. Investments in urban forestry programs, such as reforestation and community-led tree planting initiatives, are essential for improving functionality and addressing climate and health risks (FAO, 2023).

4.2.4 East Africa: Regional contributions to CO_2 absorption and air quality

Urban forests in East Africa play a crucial role in regional carbon sequestration and the enhancement of air quality. Cities such as Nairobi, Kampala, and Dar es Salaam depend on urban forests to reduce emissions and address air pollution. Nairobi's Karura Forest sequesters around 12,000 tons of CO_2 each year and filters pollutants, thereby supporting the city's sustainability objectives (Wanjiru & Matsui, 2022).
Mangrove forests along the coastline of East Africa significantly contribute to CO_2 absorption, sequestering carbon at rates five times greater than those of terrestrial forests. Coastal ecosystems play a crucial role in mitigating emissions from nearby urban centers and enhancing air quality in those regions (Maathai, 2021). Urban expansion and industrial activities pose significant threats to these forests, highlighting the need for immediate conservation efforts.

4.2.5 Tanzania: Role of urban forests in CO_2 absorption and air quality

Urban forests in Tanzania play a vital role in CO_2 absorption and enhancing air quality in the rapidly expanding cities of Dar es Salaam, Dodoma, and Arusha. Tanzania's forests sequester an estimated 1.5 billion tons of carbon, playing a crucial role in national and regional climate mitigation efforts (Ministry of Natural Resources and Tourism [MNRT], 2022).

Urban forests in Dar es Salaam, such as the Pugu and Kazimzumbwi Forest Reserves, play a critical role in reducing emissions from vehicles, industries, and households through CO_2 sequestration and pollutant filtration. These forests encounter challenges including illegal logging, urban encroachment, and insufficient policy enforcement, which diminish their capacity to address air quality and climate concerns (Nyika, 2020).

4.2.6 Kazimzumbwi Forest Reserve: A case study

Urban forests, such as the Kazimzumbwi Forest Reserve (KFR), are essential for mitigating climate change by sequestering carbon dioxide (CO_2) and enhancing urban ecological stability. KFR is projected to sequester approximately 1,296,626.72 tons of CO_2 in 2024, highlighting its critical role in urban carbon management (Pan et al., 2011). The dense forest cover, particularly mature trees, contributes significantly to this capacity, serving as a natural carbon reservoir while also providing vital ecosystem services like regulating microclimates and conserving biodiversity (Chapin et al., 2000).

The ecological implications of KFR's carbon sequestration potential are profound, especially in the face of increasing urbanization and deforestation pressures. Urban forests help mitigate localized environmental issues, such as urban heat islands and air pollution, while contributing to global climate change efforts (Seto et al., 2012). The socio-economic benefits are equally significant; the carbon sequestration potential can be monetized through carbon markets, such as the REDD+ program, representing a substantial economic asset (Angelsen, 2009). Additionally, urban forests enhance resilience to climate impacts, reducing economic costs associated with extreme weather events and improving the quality of life for urban residents (McPherson et al., 2017; Nowak et al., 2020).

However, KFR faces challenges, including deforestation, illegal logging, and encroachment for agricultural or urban development, which threaten its carbon sequestration capacity and ecological functions (FAO, 2018). Without effective management, the reserve's potential could decline, leading to increased CO_2 emissions and heightened climate risks.

To enhance KFR's carbon sequestration potential, several strategies are recommended. These include forest conservation and restoration, prioritizing native species to maintain ecological integrity (Chazdon et al., 2020); integrating KFR into carbon markets to generate revenue for conservation (Angelsen, 2009); strengthening policies against illegal activities (FAO, 2018); engaging local communities in conservation efforts to ensure sustainability (Adhikari et al., 2014); and incorporating forests into urban planning to protect green spaces (McDonald et al., 2020).

Thus, the Kazimzumbwi Forest Reserve's ability to sequester over 1.2 million tons of CO_2 underscores its vital role in urban climate mitigation and ecological sustainability. Protecting and restoring such urban forests is crucial for addressing environmental, economic, and social challenges, thereby fostering a resilient and sustainable urban future.

CHAPTER FIVE

5. CULTURAL AND SOCIAL SERVICES

5.1 Recreational, Aesthetic, and Mental Health Benefits

Urban forests serve as essential ecological resources and play a crucial role in enhancing urban quality of life. Their benefits for recreation, aesthetics, and mental health markedly improve urban living by providing areas for physical activity, social interaction, cultural engagement, and psychological well-being. This section examines the benefits from a global perspective, focusing specifically on the Kazimzumbwi Forest Reserve in Tanzania.

5.1.1 Global benefits of urban forests

Urban forests worldwide offer vital recreational and aesthetic benefits. They function as green spaces for outdoor activities, including walking, jogging, cycling, and picnicking, thereby promoting healthier lifestyles. Research conducted in urban areas such as New York and London indicates that proximity to urban forests markedly enhances physical activity levels among residents, thereby decreasing the risks of obesity, cardiovascular diseases, and diabetes (Nowak et al., 2022).

Urban forests improve the visual aesthetics of cities, playing a significant role in their cultural and historical identity. Prominent instances are Central Park in New York and Hyde Park in London, which enhance urban aesthetics and draw millions of tourists each year, thereby bolstering local economies (FAO, 2023).

Urban forests significantly contribute to mental health outcomes. Research indicates that exposure to greenery alleviates stress, anxiety, and depression, while enhancing mood and cognitive function. Research conducted in Japan demonstrates the advantages of "forest bathing" (shinrin-yoku), a practice that entails engaging with forest environments to improve mental well-being (Hansen et al., 2021). Urban forests in European cities are associated with reduced occurrences of mental health disorders, especially in regions with greater tree densities (World Health Organization [WHO], 2022).

5.1.2 Recreational and aesthetic benefits in Africa

Urban forests in Africa function as essential recreational areas within rapidly urbanizing cities. Urban green spaces in cities such as Nairobi, Lagos, and Accra facilitate physical activity, social interactions, and cultural events for residents. Karura Forest in Nairobi draws thousands of visitors each week, providing trails, picnic areas, and cultural sites that address various recreational requirements (Wanjiru & Matsui, 2022).

The aesthetic advantages of urban forests in Africa are substantial. Urban landscapes are enhanced, property values are improved, and scenic beauty is provided, fostering pride and a sense of belonging among urban residents. Trees in African cities mitigate the urban heat island effect, fostering comfortable microclimates that enhance the enjoyment of public spaces (UN-Habitat, 2022).

5.1.3 Sub-Saharan Africa: Mental health and community well-being

Sub-Saharan Africa (SSA) is experiencing increasing mental health challenges attributed to rapid urbanization, economic pressures, and socio-political instability. Urban forests in SSA are essential in mitigating these issues by creating therapeutic environments that enhance mental well-being. Research conducted in Addis Ababa and Kampala indicates that access to urban forests correlates with decreased stress and improved mental health outcomes (Myers et al., 2023). Additionally, urban forests contribute significantly to community well-being in SSA. These areas promote social cohesion by serving as venues for cultural events, education, and communal activities. Community-based conservation initiatives further reinforce these connections, as local residents collaborate to manage and restore urban forests (African Development Bank [AfDB], 2022).

5.1.4 Recreational and mental health benefits in East Africa

Urban forests in East Africa provide significant recreational and mental health advantages. Urban forests in cities such as Nairobi, Kampala, and Dar es Salaam serve as essential spaces for outdoor activities and contribute to the promotion of well-being. Karura Forest in Nairobi and Mabira Forest Reserve in Kampala serve as notable destinations for locals and tourists, providing trails, educational programs, and cultural exhibits (Wanjiru & Matsui, 2022).

The significance of mental health benefits is pronounced in East Africa, characterized by urban stressors including overcrowding, pollution, and socio-economic disparities.

Studies demonstrate that engagement with urban forests positively influences mood, decreases stress hormone levels, and bolsters psychological resilience (Nyika, 2020).

5.1.5 Tanzania: Enhancing urban well-being through forests

Urban forests in Tanzania play a crucial role in enhancing recreational, aesthetic, and mental health benefits, especially in the rapidly expanding cities of Dar es Salaam, Dodoma, and Arusha. Forests serve as vital green spaces for residents, facilitating physical activity, social interaction, and a connection with nature (Ministry of Natural Resources and Tourism [MNRT], 2022).

The aesthetic advantages of urban forests in Tanzania encompass the improvement of cityscape visual appeal and the stimulation of tourism. The Pugu and Kazimzumbwi Reserves near Dar es Salaam are recognized for their natural beauty, drawing visitors for leisure and ecotourism activities. These forests enhance microclimates, contributing to more comfortable and sustainable urban living (Mgaya & Samwel, 2021).

Urban forests in Tanzania significantly contribute to mental health. They offer therapeutic environments that mitigate the stresses associated with urban living, especially in highly populated regions. Research suggests that engagement with urban forests is associated with decreased anxiety and depression, thereby enhancing mental health outcomes for urban populations (Nyika, 2020).

5.1.6 Kazimzumbwi Forest Reserve: A case study

5.1.6.1 Data used and methods

Figure 5.1 below shows the flow chart of the methodological approach used in this study for the estimation of the ecosystem service values (ESVs) for 1994, 2004, and 2024 years and the computation of changes between studies periods.

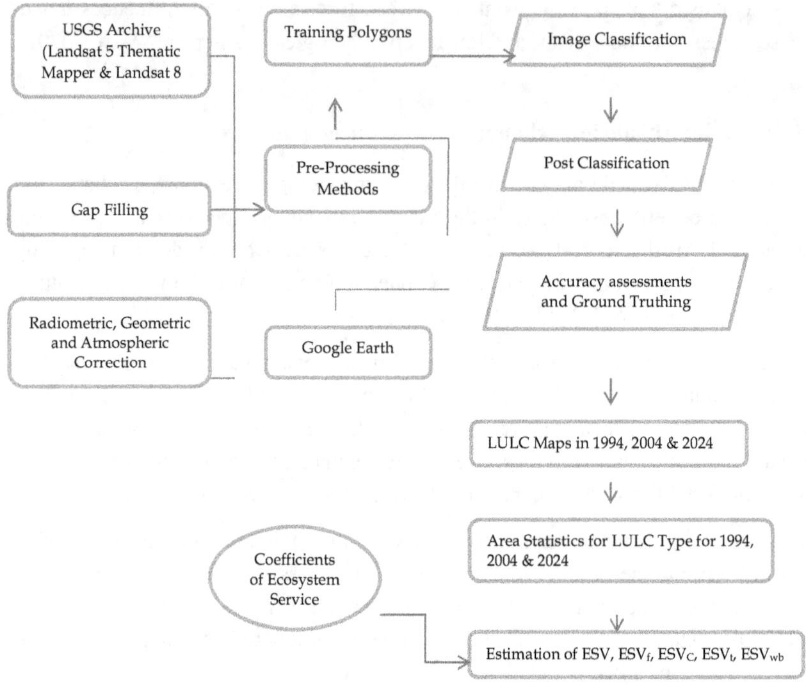

Figure 5.1: Flowchart of the methodological approach for this study

The LULC datasets in Table 2.2 and biome equivalents with their corresponding ecosystem service value coefficients (VC) in 1994 US\$ ha^{-1}year^{-1} for local and global VC shown in Table 5.1 as adapted from Kindu *et al.* (2016); Constaza *et al.* (1997 &2014); and Zella (2020).

Table 5.1: Land use and land cover (LULC) types and biome equivalents with their corresponding ecosystem service value coefficients (VC)

| LULC Type | Year & Area (ha) | | | Equivalent Biome | Local (VC) 1994 US$ ha⁻¹year⁻¹ | Global (VC) 1994 US$ ha⁻¹year⁻¹ |
	1994	2004	2024		a	b
Closed forest	2722.28	1577.58	886.64	Tropical Forest	987	2008
Open woodland	1399.65	601.068	1390.09	Tropical Forest	987	2008
Shrubs	372.28	791.796	1160.98	Tropical Forest	987	244
Grassland	235.89	1666.42	925.22	Grasslands	293	244
Bare land	245.07	338.298	540.01	Soil	0	0
Built-up	0.00	0	72.23	Urban	0	0

This study employed the benefit transfer approach to estimate economic values of ecosystem services based on the adapted local and global VC of the ecosystem services for the targeted LULC types. Detailed ecosystem service functions and their global and modified local value coefficients of each LULC type are shown in Tables 5.2 & 5.3 below as adapted from Zella (2020) and Constaza *et al.* (1997 &2014).

Table 5.2: Details of the ecosystem service functions and their modified local value coefficients for each LULC type (adapted from Zella., 2020)

| Ecosystem Services | Each LULC Types of Ecosystem Service Values (1994 US$ ha⁻¹year⁻¹) | | | |
	Closed forest	Open woodland	Shrubs	Grassland
Provisioning services:				
Water supply	8	8	8	
Food production	32	32	32	117.45
Raw material	51.2	51.2	51.2	
Genetic resources	41	41	41	
Medical services				

Sub-total	**132.2**	**132.2**	**132.2**	**117.45**
Regulating services:				
Water regulation	6	6	6	3
Waste treatment	136	136	136	87
Erosion control	245	245	245	29
Climate regulation	223	223	223	
Biological control				23
Gas regulation	13.68	13.68	13.68	7
Disturbance regulation	5	5	5	
Sub-total	**628.68**	**628.68**	**628.68**	**149**
Supporting services:				
Nutrient cycling	184.4	184.4	184.4	
Pollination	7.27	7.27	7.27	25
Soil formation	10	10	10	1
Habitat/refugia	17.3	17.3	17.3	
Sub-total	**218.97**	**218.97**	**218.97**	**26**
Cultural services:				
Recreation	4.8	4.8	4.8	0.8
Cultural	2	2	2	
Sub-total	**6.8**	**6.8**	**6.8**	**0.8**
Grand-total	**986.69**	**986.69**	**986.69**	**293.25**

Table 5.3: Details of the ecosystem service functions and their global value coefficients for each LULC type (adapted from Constaza *et al.*, 1997)

	Each LULC Types of Ecosystem Service Values (1994 US$ ha^{-1}year^{-1})			
Ecosystem Services	**Closed woodland**	**Open woodland**	**Bushland**	**Grassland**
Provisioning services:				
Water supply	8	8	8	

Food production	32	32	32	67
Raw material	315	315	315	106
Genetic resources	41	41	41	
Medical services				
Sub-total	**396**	**396**	**396**	**173**
Regulating services:				
Water regulation	6	6	6	3
Waste treatment	87	87	87	87
Erosion control	245	245	245	29
Climate regulation	223	223	223	
Biological control				23
Gas regulation				7
Disturbance regulation	5	5	5	
Sub-total	**566**	**566**	**566**	**149**
Supporting services:				
Nutrient cycling	922	922	922	
Pollination				25
Soil formation	10	10	10	1
Habitat/refugia				
Sub-total	**932**	**932**	**932**	**26**
Cultural services:				
Recreation	112	112	112	2
Cultural	2	2	2	
Sub-total	**114**	**114**	**114**	**2**
Grand-total	**2008**	**2008**	**2008**	**350**

5.1.6.2 Data analysis

To determine changes of economic values of ecosystem services resulted from LULCC of Kazimzumbwi Forest Reserve from 1994 to 2024

The LULC datasets shown in Table 5.1 used and the total value of ecosystem services in the study area for 1994, 2004 and 2024 was calculated by multiplying the area of a given LULC type by the corresponding modified ecosystem service value coefficients that were extracted from weight factors of the ecosystem services per hectare of each biome, see equation (1) adapted from Msofe *et al.*, 2020 and Constaza *et al.*, 1997 & 2014 as follows:

$$ESV = \sum_{k=0}^{k}(Ak + VCk) \dots\dots\dots\dots\dots\dots\dots\dots\dots\dots\dots\dots\dots(1)$$

where ESV = the total estimated ecosystem service value, A_k = the area (ha) and VC_k = the value coefficient (US\$ ha^{-1} year^{-1}) for LULC type 'k'. The ESVs for all land use and land cover (LULC) types were calculated. Besides, the change in the ESVs was determined by calculating the differences between the estimated values for each LULC category in 1994, 2004 and 2024. The percentage changes in the ESVs between the years were calculated based on the equation below:

$$Percentange\ ESV = \frac{(ESVt_2 - ESVt_1)}{ESVt_1} \times 100 \dots\dots\dots\dots\dots\dots\dots\dots\dots\dots(2)$$

where ESV_{t2} (US\$ ha^{-1} year^{-1}) = the estimated ecosystem service value in the most recent year, and ESV_{t1} (US\$ ha^{-1} year^{-1}) = the estimated ecosystem service value in the previous year. Positive values suggest an increase in the ESVs, whereas negative values imply a decrease in the ESVs.

To analyse changes of economic values of ecosystem functions based on LULC type of Kazimzumbwi Forest Reserve from 1994 to 2024

Estimated values of the services provided by individual ecosystem functions within the study area using the following equation:

$$ESVf = \sum_{k=0}^{k}(Ak * VC_{fk}) \dots\dots\dots\dots\dots\dots\dots\dots\dots\dots\dots\dots\dots(3)$$

where ESV_f is the estimated ecosystem service value of function f, Ak is the area (ha) and VC_{fk} is the value coefficient of the function (US\$ ha^{-1} year^{-1}) for LULC category 'k'. The contributions of the individual ecosystem functions to the overall value of the ecosystem services per year were calculated and summarized in the tables.

5.1.6.3 Results and Discussions

Changes of economic values of ecosystem services resulted from LULCC of Kazimzumbwi Forest Reserve from 1994 to 2024

Ecosystem services are critical for maintaining ecological balance and enhancing human well-being, as exemplified by the Kazimzumbwi Forest Reserve (KFR). This reserve provides essential services such as carbon sequestration, water regulation, biodiversity conservation, and cultural benefits. However, land use and land cover (LULC) changes have significantly impacted the economic valuation of these services over time. An analysis of KFR's ecosystem services from 1994 to 2024 reveals a concerning decline in their total economic value, estimated at approximately US$ 216 million, or 18%, primarily due to forest cover loss and ecosystem degradation (Costanza et al., 1997; Pan et al., 2011).

In 1994, KFR's ecosystem services were at their peak, reflecting a relatively intact forest ecosystem. By 2004, early signs of deforestation began to emerge, leading to moderate declines in service values. Projections for 2024 indicate a significant reduction, driven by ongoing deforestation, urban expansion, and agricultural encroachment. This trend highlights the vulnerability of ecosystem services to human activities and the associated economic costs of forest degradation (McDonald et al., 2020).

The decline in ecosystem service values is closely tied to LULC changes, particularly the conversion of forests to agricultural and urban areas. Deforestation and fragmentation are major contributors to this loss, as forests are vital for carbon storage, water regulation, and biodiversity (IPCC, 2021). Urbanization, while economically beneficial, often leads to environmental challenges that further diminish ecosystem services.

The estimated loss of US$ 216 million underscores the economic implications of unsustainable land-use practices. This figure reflects the opportunity costs associated with the degradation of services such as carbon sequestration, which exacerbates climate change, and water regulation, which increases flooding risks and water scarcity (Chapin et al., 2000). Additionally, biodiversity loss undermines ecosystem resilience and agricultural productivity (Pimm et al., 2014).

Local communities, particularly marginalized populations, are disproportionately affected by the decline in ecosystem services, which exacerbates socio-economic vulnerabilities. The loss of resources such as fuelwood and clean water diminishes their quality of life and access to essential services (FAO, 2018).

To counteract these trends, strategies such as integrating ecosystem services into economic planning, promoting forest restoration, engaging communities in conservation, and implementing Payment for Ecosystem Services (PES) schemes are essential. Urban green infrastructure can also mitigate the impacts of land-use changes (Chazdon et al., 2020; Engel et al., 2008). Urgent action is needed to preserve KFR's ecological and socio-economic functions, ensuring long-term benefits for both urban and rural communities.

Table 5.4: Local ecosystem services values (ESV) year⁻¹distribution between 1994 and 2024

LULC	1994		2004		2024	
	(ESV)	(%)	(ESV)	(%)	(ESV)	(%)
Closed forest	2,686,890.4	59.6	1,557,071.5	45.5	875,113.7	23.9
Open woodland	1,381,454.6	30.7	593,254.1	17.3	,372,018.8	37.4
Shrubs	367,440.4	8.2	781,502.7	22.9	1,145,887.3	31.3
Grassland	69,115.8	1.5	488,261.1	14.3	271,089.5	7.4
Bare land	-	-	-	-	-	0.0
Built-up	-	-	-	-	-	0.0
Total	**4,504,901.0**	**100.0**	**3,420,089.3**	**100.0**	**3,664,109.2**	**100.0**

Ecosystem services (ES) are vital for ecological integrity, livelihoods, and human well-being, yet their economic valuation varies significantly based on methodology and context. A study on the Kazimzumbwi Forest Reserve (KFR) reveals a dramatic decline in global economic values of ecosystem services (ESVs) over three decades, with a 41% decrease (approximately US\$ 612 million) from 1994 to 2024. This decline is attributed to the degradation and conversion of high-value land use types, such as forests, into less valuable uses like agriculture and urban development (Costanza et al., 1997; de Groot et al., 2012).

The findings indicate a consistent disparity between global and local ESVs across the years studied. In 1994, global ESVs were 20.4% higher than local values, increasing to 26.2% in 2004, before narrowing to 14.4% by 2024. Notably, global ESVs underperformed in certain categories, highlighting the challenges of applying global valuation frameworks to local contexts where socio-economic and ecological dynamics differ (Adhikari et al., 2014).

The divergence between global and local ESVs raises critical questions for ecosystem management. Global ESVs, derived from standardized coefficients, may not accurately reflect local values, leading to potential mismanagement. Effective ecosystem management must balance ecological benefits with local community needs, as over-reliance on global ESVs risks marginalizing stakeholders who depend on ecosystem resources (IPCC, 2021).

The decline in ESVs is driven by factors such as deforestation, urbanization, and resource overexploitation. Deforestation for agriculture and urban development diminishes the capacity of ecosystems to provide essential services like carbon sequestration and biodiversity support (Seto et al., 2012; McDonald et al., 2020).

Despite the importance of local ESVs, global ESVs offer a broader perspective on the value of ecosystem services, emphasizing their contributions to global climate regulation and biodiversity conservation (Chapin et al., 2000; Costanza et al., 2014). Therefore, integrating local and global perspectives in valuation frameworks is crucial. Strategies such as stakeholder engagement, payment for ecosystem services (PES), and aligning policies with global environmental goals can help balance socio-economic development with ecological conservation (Engel et al., 2008; UN, 2015).

Thus, the significant decline in KFR's ESVs underscores the complexities of valuing ecosystem services and the need for integrated approaches that prioritize natural capital and stakeholder involvement to ensure sustainable management for future generations.

Table 5.5: Global Ecosystem Services Values (ESV) distribution between 1994 and 2024

LULC	1997		2005		2016	
	(ESV)	(%)	(ESV)	(%)	(ESV)	(%)
Closed forest	5,466,338.2	64.9	3,167,780.6	63.7	1,780,373.1	35.0
Open woodland	2,810,497.2	33.4	1,206,944.5	24.3	2,791,300.7	54.9
Shrubs	90,836.3	1.1	193,198.2	3.9	283,279.1	5.6
Grassland	57,557.2	0.7	406,606.5	8.2	225,753.7	4.4
Bare land	-	-	-	-	-	0.0
Built-up	-	-	-	-	-	0.0
Total	**8,425,228.9**	**100.0**	**4,974,529.9**	**100.0**	**5,080,706.6**	**100.0**

Changes of economic values of ecosystem services of LULCC biomes of Kazimzumbwi forest reserve from 1994 to 2024

The economic value of ecosystem services (ESV) is crucial for understanding the benefits derived from various land-use and land-cover (LULC) types, particularly in the Kazimzumbwi Forest Reserve (KFR). An analysis of ESV changes from 1994 to 2024 indicates significant trends influenced by shifts in LULC biomes. The study employed both local and global valuation coefficients, revealing notable variations in ESV changes and their implications for policy development.

From 1994 to 2004, total ESV decreased significantly, with local coefficients showing a decline of US$ 1.08 million and global coefficients indicating a larger drop of US$

3.45 million. This decline was attributed to the degradation of high-value biomes, particularly closed forests and open woodlands. Conversely, between 2004 and 2024, ESV showed a slight recovery, with local coefficients indicating an increase of US$ 0.24 million and global coefficients reflecting a smaller rise of US$ 0.11 million. This recovery is linked to stabilization in some biomes and increases in shrub and grassland cover, although these biomes provide fewer ecosystem services compared to closed forests (Costanza et al., 1997; Chapin et al., 2000).

The annual rate of ESV change from 1994 to 2024 was estimated at US$ 0.84 million using local coefficients and US$ 3.35 million with global coefficients, highlighting a significant discrepancy of 74.9% (US$ 2.50 million). This suggests that global coefficients are more sensitive to capturing the macroeconomic values of ecosystem services. The degradation of closed forests and open woodlands was particularly pronounced, with local coefficients indicating a 215.5% decline in closed forests and a 1.1% decline in open woodlands.

The implications of these findings stress the need for integrated valuation frameworks that reconcile local and global perspectives. The differences in valuation methodologies underscore the importance of using global coefficients to inform macroeconomic policies and international conservation commitments, such as the United Nations Sustainable Development Goals (UN, 2015). Policymakers should prioritize high-value ecosystems and incorporate ESV trends into decision-making models, addressing uncertainties in ecosystem service management (Bateman et al., 2013).

Thus, the study highlights significant changes in ESV in KFR from 1994 to 2024, driven by land-cover transitions and valuation methodology disparities. While there was a substantial decline in ESV from 1994 to 2004, a modest recovery occurred from 2004 to 2024. Integrating global valuation frameworks into national policies is essential for sustainable ecosystem management, balancing ecological conservation with socio-economic development (IPCC, 2021).

Table 5.6: Changes in Local ESV from 1994 to 2024

LULC	1994 – 2024			2004 – 2024		
	Change in ESV (US$))	% change	Annual Rate of Change (ESV/year) (US$))	Change in ESV (US$))	% change	Annual Rate of Change (ESV/year) (US$))
Closed forest	1,129,818.9	104.1	112,981.9	681,957.8	(279.5)	136,391.6
Open woodland	788,200.4	72.7	78,820.0	(778,764.7)	319.1	(155,752.9)
Shrubs	(414,062.3)	(38.2)	(41,406.2)	(364,384.6)	149.3	(72,876.9)
Grassland	(419,145.3)	(38.6)	(41,914.5)	217,171.6	(89.0)	43,434.3
Bare land	-	-	-	-	-	-
Built-up	-	-	-	-	-	-
Total	1,084,811.8	100.0	108,481.2	(244,019.9)	100.0	(48,804.0)

Table 5.7: Changes in Global ESV from 1994 to 2024

LULC	1994 – 2004			2004 – 2024		
	Change in ESV (US$))	% change	Annual Rate of Change (ESV/year) (US$))	Change in ESV (US$))	% change	Annual Rate of Change (ESV/year) (US$))
Closed forest	2,298,557.6	66.6	229,855.8	1,387,407.5	(1,306.7)	277,481.5
Open woodland	1,603,552.7	46.5	160,355.3	(1,584,356.2)	1,492.2	(316,871.2)
Shrubs	(102,361.9)	(3.0)	(10,236.2)	(90,080.9)	84.8	(18,016.2)
Grassland	(349,049.3)	(10.1)	(34,904.9)	180,852.8	(170.3)	36,170.6
Bare land	-	-	-	-	-	-
Built-up	-	-	-	-	-	-
Total	**3,450,699.0**	**100.0**	**345,069.9**	**(106,176.8)**	**100.0**	**(21,235.4)**

Changes of economic values of ecosystem functions based on LULC Type of Kazimzumbwi forest reserve from 1994 to 2024

The Kazimzumbwi Forest Reserve (KFR) has experienced significant losses in ecosystem functions from 1994 to 2024, primarily due to anthropogenic activities such as deforestation, agriculture, and urban expansion. The economic valuation of these functions reveals a loss of approximately US$ 903,291.2 using local coefficients and US$ 1,880,186.6 using global coefficients, indicating severe macroeconomic impacts from ecosystem degradation (Chapin et al., 2000; Costanza et al., 1997). Closed forests accounted for over 50% of the losses, while open woodlands contributed around 26%.

The findings indicate that while provisioning services, such as timber and fuelwood extraction, have seen gains, supporting services (e.g., soil formation, nutrient cycling) and regulatory services (e.g., carbon sequestration) have suffered significant degradation. This decline undermines the ecological stability of KFR, affecting biodiversity and the reserve's ability to mitigate climate change (Pan et al., 2011). The economic implications are profound, as the losses reflect missed opportunities in carbon markets and sustainable forest management, emphasizing the need for a balance between short-term resource extraction and long-term ecological health (de Groot et al., 2012).

Socio-economic drivers, particularly population growth and reliance on forest resources, have accelerated this degradation, leading to a cycle of poverty and environmental decline (FAO, 2018). The diminishing cultural services of KFR further exacerbate the situation, reducing its value as a recreational and cultural site, crucial for urban populations (McDonald et al., 2020).

To combat these challenges, integrated management strategies are essential. These include large-scale restoration programs for degraded forests (Chazdon et al., 2020), sustainable resource management practices, and alternative livelihood development to reduce dependency on forest exploitation (Adhikari et al., 2014). Additionally, incorporating ecosystem service valuation into urban planning can help mitigate the impacts of urbanization (Seto et al., 2012).

Global valuation frameworks, such as REDD+ and Payment for Ecosystem Services (PES), can provide necessary funding and incentives for conservation efforts (Angelsen, 2009; Engel et al., 2008). Raising awareness about the importance of ecosystem services and engaging local communities in management practices are also critical for fostering sustainable use (FAO, 2018).

Thus, the significant decline in KFR's ecosystem functions underscores the urgent need for integrated strategies that prioritize ecological conservation alongside socio-economic development, leveraging global frameworks to enhance conservation efforts (Costanza et al., 2014).

Table 5.8: Local economic values of ecosystem functions (US$) from 1994 to 2024

LULC	Ecosystem services	1994	2004	2024	Relative change
	Provisioning services	359,885.4	208,556.1	117,213.8	242,671.6
	Regulating services	1,711,443.0	991,793.0	557,412.8	1,154,030.2
	Supporting services	596,097.7	345,442.7	194,147.6	401,950.1
Closed	Cultural services	18,511.5	10,727.5	6,029.2	12,482.3
forest	**Sub-total**	**2,685,937.6**	**1,556,519.3**	**874,803.4**	**1,811,134.2**
	Provisioning services	185,033.7	79,461.2	183,769.9	1,263.8
	Regulating services	879,932.0	377,879.4	873,921.8	6,010.2
	Supporting services	306,481.4	131,615.9	304,388.0	2,093.4
Open	Cultural services	9,517.6	4,087.3	9,452.6	65.0
woodland	**Sub-total**	**1,380,964.7**	**593,043.8**	**1,371,532.3**	**9,432.4**
	Provisioning services	49,215.4	104,675.4	153,481.6	(104,266.1)
	Regulating services	234,045.0	497,786.3	729,884.9	(495,839.9)
	Supporting services	81,518.2	173,379.6	254,219.8	(172,701.6)
	Cultural services	2,531.5	5,384.2	7,894.7	(5,363.2)
Shrub	**Sub-total**	**367,310.1**	**781,225.5**	**1,145,481.0**	**(778,170.9)**
	Provisioning services	27,705.3	195,721.0	108,667.1	(80,961.8)
	Regulating services	35,147.6	248,296.6	137,857.8	(102,710.2)
	Supporting services	69,174.7	43,326.9	24,055.7	45,119.0
	Cultural services	188.7	1,333.1	740.2	(551.5)
Grassland	**Sub-total**	**132,216.3**	**488,677.6**	**271,320.8**	**(139,104.5)**
GRAND TOTAL		**4,566,428.6**	**3,419,466.2**	**3,663,137.4**	**903,291.2**

Table 5.9: Global economic values of ecosystem functions (US$) from 1994 to 2024

LULC	Ecosystem services	1994	2004	2024	Relative change
	Provisioning services	1,078,022.9	624,721.7	351,109.4	726,913.5
	Regulating services	1,540,810.5	892,910.3	501,838.2	1,038,972.3
	Supporting services	2,537,165.0	1,470,304.6	826,348.5	1,710,816.5
Closed	Cultural services	310,339.9	179,844.1	101,077.0	209,262.9
woodland	**Sub-total**	**5,466,338.3**	**3,167,780.7**	**1,780,373.1**	**3,685,965.2**
	Provisioning services	554,261.4	238,022.9	550,475.6	3,785.8
	Regulating services	792,201.9	340,204.5	786,790.9	5,411.0
	Supporting services	1,304,473.8	560,195.4	,295,563.9	8,909.9
Open	Cultural services	159,560.1	68,521.8	158,470.3	1,089.8
woodland	**Sub-total**	**2,810,497.2**	**1,206,944.6**	**2,791,300.7**	**19,196.5**
	Provisioning services	147,422.9	313,551.2	459,748.1	(312,325.2)
	Regulating services	210,710.5	448,156.5	657,114.7	(446,404.2)
	Supporting services	346,965.0	737,953.9	1,082,033.4	(735,068.4)
	Cultural services	42,439.9	90,264.7	132,351.7	(89,911.8)
Bushland	**Sub-total**	**747,538.3**	**1,589,926.3**	**2,331,247.9**	**(1,583,709.6)**
	Provisioning services	40,809.0	288,290.7	160,063.1	(119,254.1)
	Regulating services	35,147.6	248,296.6	137,857.8	(102,710.2)
	Supporting services	6,133.1	43,326.9	24,055.7	(17,922.6)
	Cultural services	471.8	3,332.8	1,850.4	(1,378.6)
Grassland	**Sub-total**	**82,561.5**	**583,247.0**	**323,827.0**	**(241,265.5)**
GRAND TOTAL		**9,106,935.3**	**6,547,898.6**	**7,226,748.7**	**1,880,186.6**

5.2 Community Involvement in Forest Preservation

Community engagement is fundamental to successful forest conservation. Local communities rely on forests for livelihoods, cultural practices, and ecological services, positioning them as essential stakeholders in conservation initiatives. This section examines community involvement in forest preservation, focusing on a global context and specifically narrowing down to Africa, Sub-Saharan Africa, East Africa, Tanzania, and the Kazimzumbwi Forest Reserve.

5.2.1 Global perspectives on community involvement in forest preservation

The effectiveness of forest preservation initiatives worldwide is increasingly linked to community involvement. The transition from centralized, top-down conservation strategies to community-based models has increased, acknowledging the essential contribution of local knowledge and stewardship in the management of forest resources (Food and Agriculture Organization [FAO], 2023).

Community Forest Management (CFM) programs in nations such as Nepal, Mexico, and India illustrate the effectiveness of local involvement in promoting forest conservation. In Nepal, CFM initiatives have resulted in notable enhancements in forest cover and biodiversity, concurrently offering economic advantages to local communities (Chhetri et al., 2022). Mexico's ejido system enables communities to sustainably manage and profit from their forest resources, demonstrating the effectiveness of community engagement in meeting conservation objectives (Bray, 2021).

Community-based approaches, while beneficial, encounter challenges including inadequate funding, disputes regarding resource utilization, and restricted enforcement capacity. Integrating traditional knowledge with modern conservation techniques has been effective in addressing these issues and promoting sustainable forest management worldwide (UNEP, 2022).

5.2.2 Community involvement in forest preservation in Africa

In Africa, forests play a critical role in supporting livelihoods, maintaining biodiversity, and regulating climate. Therefore, community engagement is essential for tackling deforestation and degradation. Participatory Forest Management (PFM) programs are being increasingly implemented throughout the continent, engaging communities in decision-making, monitoring, and benefit-sharing (African Forest Forum, 2022).

Ethiopia, Ghana, and Kenya have successfully implemented public financial management initiatives. The Bale Mountains Eco-Region REDD+ Project in Ethiopia involves local communities in forest conservation and offers alternative livelihoods, including beekeeping and eco-tourism (Amente et al., 2023). Kenya's Green Belt Movement, established by Wangari Maathai, has engaged communities in the planting of millions of trees, thereby fostering forest restoration and enhancing climate resilience (Maathai, 2021).

Challenges in Africa encompass land tenure insecurity, inadequate institutional frameworks, and conflicts over resources. To address these issues, it is essential to implement robust policies, enhance capacity-building, and establish equitable benefit-sharing mechanisms that encourage community participation in forest preservation (FAO, 2023).

5.2.3 Sub-Saharan Africa: Engaging communities in forest conservation

Sub-Saharan Africa (SSA) contains significant forest ecosystems, notably the Congo Basin and miombo woodlands. Community engagement in forest preservation is essential for maintaining these ecosystems, which are under considerable threat from logging, agriculture, and urban development (Global Forest Watch, 2022).

CFM initiatives in SSA frequently highlight the incorporation of traditional ecological knowledge alongside scientific methodologies. The Collaborative Forest Management (CFM) model in Uganda facilitates community co-management of forest reserves alongside government agencies, leading to enhanced forest cover and diminished conflicts (Myers et al., 2023).

Community-based programs in Sub-Saharan Africa frequently encounter challenges related to funding limitations and insufficient technical support. Enhancing collaboration among governments, non-governmental organizations (NGOs), and local communities is crucial for tackling these challenges and improving the efficacy of forest preservation initiatives (African Development Bank [AfDB], 2022).

5.2.4 Community involvement in East Africa

Community participation is essential to forest conservation initiatives in East Africa, especially in Kenya, Uganda, and Tanzania. Participatory approaches have demonstrated efficacy in mitigating deforestation and advancing sustainable forest management practices.

The Green Belt Movement in Kenya exemplifies community-driven reforestation efforts. The movement has restored degraded landscapes, enhanced community resilience, and created income-generating opportunities through the mobilization of local women's groups (Maathai, 2021). The Mabira Forest Integrated Community Organization in Uganda collaborates with local communities to safeguard the Mabira Forest, while also offering education and alternative livelihoods (Nyika, 2020).

Despite these achievements, challenges including land-use conflicts, insufficient funding, and ineffective enforcement mechanisms continue to exist. Enhancing community ownership and incorporating conservation into wider development agendas can effectively tackle these challenges and promote the sustainability of forest preservation initiatives in East Africa (Wanjiru & Matsui, 2022).

5.2.5 Community-based forest preservation in Tanzania

Tanzania has adopted community-based strategies for forest conservation, exemplified by initiatives such as Participatory Forest Management (PFM) and Joint Forest Management (JFM). These programs enable local communities to oversee forest resources, distribute benefits, and engage in decision-making processes.

Public Financial Management initiatives in Tanzania, particularly in the Eastern Arc Mountains, have led to improvements in forest governance, a reduction in illegal logging, and advancements in biodiversity conservation. Communities participating in these programs gain advantages through capacity-building, financial incentives, and enhanced access to resources such as non-timber forest products (Ministry of Natural Resources and Tourism [MNRT], 2022).

Challenges including limited funding, inadequate enforcement of conservation laws, and conflicts over resource utilization impede the full potential of community-based forest preservation. To address these challenges, it is essential to implement stronger policy frameworks, increase investment in capacity-building, and enhance community engagement (Mgaya & Samwel, 2021).

5.2.6 Kazimzumbwi Forest Reserve: Community involvement in preservation

The Kazimzumbwi Forest Reserve, situated near Dar es Salaam, underscores the significance of community engagement in forest conservation. Kazimzumbwi, one of Tanzania's oldest coastal forests, is threatened by urban encroachment, illegal logging, and agricultural expansion. Community participation is essential for tackling these challenges and revitalizing the forest's ecological and socio-economic value (Mwakalila et al., 2023).

S/n	Community involvement in preservation	Explanation
1.	Community-based initiatives	Communities in the Kazimzumbwi area participate in reforestation, agroforestry, and conservation education initiatives. The initiatives seek to rehabilitate degraded regions, increase forest cover, and improve the livelihoods of communities. Tree-planting campaigns and beekeeping projects generate income and promote conservation (Nyika, 2020).
2.	Challenges	Kazimzumbwi encounters challenges including insufficient funding, ineffective enforcement of conservation policies, and competing land uses. Resolving these issues necessitates enhanced collaboration among local communities, government agencies, and NGOs (MNRT, 2022).
3.	Opportunities	Incorporating Kazimzumbwi into the urban planning framework of Dar es Salaam and fostering eco-tourism may improve community involvement and offer sustainable financial support for conservation efforts. Moreover, utilizing traditional knowledge and enhancing community capacity can fortify preservation initiatives and guarantee long-term sustainability (Mgaya & Samwel, 2021).

PART III: CHALLENGES AND SOLUTIONS

CHAPTER SIX

6. THREATS TO URBAN FORESTS

6.1 Climate Change Impacts

Climate change represents a major challenge for ecosystems and human societies in the 21st century. Urban forests are essential elements of ecological systems, influencing and being influenced by climate change impacts. This section examines the impacts of climate change on forests and their ecological and socio-economic functions, starting with a global overview and focusing on Africa, Sub-Saharan Africa, East Africa, Tanzania, and ultimately, the Kazimzumbwi Forest Reserve.

6.1.1 Global climate change impacts on forests

Climate change exerts significant impacts on forests worldwide, modifying their structure, function, and distribution. Increasing temperatures, altered precipitation patterns, and a higher incidence of extreme weather events are disrupting forest ecosystems. Research demonstrates that extended periods of drought and heatwaves lead to significant tree mortality, resulting in decreased forest cover and biodiversity (Allen et al., 2015).

Forests are influenced by secondary effects of climate change, including heightened pest outbreaks and wildfires. In North America, rising temperatures have enabled the proliferation of bark beetles, resulting in the destruction of millions of hectares of pine forests (Seidl et al., 2017). The Amazon rainforest is facing diminished resilience as a result of extended droughts, jeopardizing its function as a global carbon sink (Phillips et al., 2019).

Urban forests encounter heightened pressures, notably the urban heat island effect, which exacerbates local temperature increases. The challenges posed jeopardize the capacity of forests to deliver essential ecosystem services, including carbon sequestration, temperature regulation, and biodiversity conservation (Nowak et al., 2022).

6.1.2 Climate change impacts in Africa

Africa experiences significant impacts from climate change, even though its contribution to global greenhouse gas emissions is minimal. Increasing temperatures, modified precipitation patterns, and a higher incidence of droughts and floods are significantly affecting forest ecosystems throughout the continent (African Development Bank [AfDB], 2022).

African forests exhibit significant vulnerability to climate variability, impacting biodiversity, livelihoods, and carbon storage. The Congo Basin, the second-largest rainforest globally, is threatened by extended dry periods and deforestation, which diminish its ability to sequester carbon and sustain wildlife (Global Forest Watch, 2022).

Urban forests in African cities play a vital role in mitigating the impacts of climate change, yet they face growing threats. Urban forests in cities such as Nairobi and Accra are under stress due to increasing temperatures, water scarcity, and urban expansion, which diminishes their ecological and socio-economic benefits (Wanjiru & Matsui, 2022).

6.1.3 Sub-Saharan Africa: Forests and climate change

Sub-Saharan Africa (SSA) exhibits significant vulnerability to climate change impacts, attributed to its reliance on natural resources and constrained adaptive capacity. Forest ecosystems in Sub-Saharan Africa are undergoing changes in species composition and distribution as a result of increasing temperatures and altered rainfall patterns (Niang et al., 2014).

Miombo woodlands, prevalent in extensive regions of Sub-Saharan Africa, exhibit significant vulnerability to climate-induced droughts and fire occurrences. The alterations impact biodiversity and the livelihoods of millions reliant on these forests for fuel, food, and income (Chidumayo & Gumbo, 2010).

Urban forests in Sub-Saharan Africa are essential for mitigating climate change effects through shade provision, urban cooling, and air pollution reduction. Urbanization and poor management compromise their resilience to climate change, highlighting the need for immediate conservation initiatives (FAO, 2023).

6.1.4 East Africa: Regional climate change impacts

East Africa is undergoing elevated temperatures and heightened climate variability, which are substantially impacting its forests. Extended droughts, irregular precipitation, and an increase in extreme weather occurrences are leading to forest degradation and a decline in biodiversity (Nyika, 2020).

Mangrove forests along the coastline of East Africa exhibit significant vulnerability to climate change. Rising sea levels and heightened storm surges pose significant risks to essential ecosystems that function as carbon sinks and safeguard coastal communities against erosion and flooding (Maathai, 2021).

Urban forests in East African cities such as Kampala, Nairobi, and Dar es Salaam encounter comparable challenges. They experience stress from climate-induced factors, including heatwaves and water scarcity, as well as anthropogenic pressures such as deforestation and land-use changes (Wanjiru & Matsui, 2022).

6.1.5 Climate change impacts in Tanzania

The forests of Tanzania are profoundly impacted by climate change, resulting in extensive consequences for biodiversity, livelihoods, and urban resilience. Increasing temperatures and altered precipitation patterns are leading to changes in forest ecosystems, diminishing their ability to deliver ecosystem services (Ministry of Natural Resources and Tourism [MNRT], 2022).

Coastal forests, including the Pugu and Kazimzumbwi Reserves, exhibit significant vulnerability to the impacts of climate change. Extended drought conditions and elevated temperatures diminish tree growth and resilience, whereas heightened storm intensity jeopardizes forest stability and biodiversity (Mgaya & Samwel, 2021).

Urban forests in Tanzania, especially in rapidly expanding cities such as Dar es Salaam and Dodoma, encounter heightened pressures due to urbanization. Forests play a crucial role in mitigating urban heat islands and sequestering carbon; however, they are facing increasing degradation from encroachment and climate stressors (Nyika, 2020).

6.1.6 Kazimzumbwi Forest Reserve: Climate change impacts

The Kazimzumbwi Forest Reserve, located near Dar es Salaam, serves as a critical ecological buffer against the impacts of climate change. However, it is not immune to these effects.

S/n	Climate change impacts	Explanation
1.	Biodiversity loss	Climate change is inducing alterations in species composition in Kazimzumbwi, posing risks to endemic and rare species. Increased temperatures and decreased precipitation impact the growth and survival of indigenous tree species, thereby diminishing forest resilience and biodiversity (Mwakalila et al., 2023).
2.	Decreased carbon sequestration	Kazimzumbwi's function as a carbon sink is undermined by climate-related stressors, including extended droughts and heatwaves. The conditions diminish the forest's capacity to absorb and store carbon, thereby compromising its role in climate mitigation (Mgaya & Samwel, 2021).
3.	Increased vulnerability to fires	In Kazimzumbwi, drier conditions and elevated temperatures heighten the risk of wildfires. Fires result in the destruction of vegetation and the release of stored carbon into the atmosphere, thereby exacerbating the impacts of climate change (Nyika, 2020).
4.	Community impacts	Local communities reliant on Kazimzumbwi for resources, including fuelwood and non-timber forest products, experience diminished availability as a result of climate-induced forest degradation. This endangers their livelihoods and intensifies susceptibility to climate impacts (MNRT, 2022).
5.	Conservation challenges and opportunities	Mitigation efforts for climate change impacts in Kazimzumbwi encompass reforestation initiatives, community-based conservation strategies, and the incorporation of forest considerations into urban planning frameworks. The initiatives seek to improve the forest's resilience and restore its ecological and socio-economic functions (Mwakalila et al., 2023).

6.2 Anthropogenic Pressures and Deforestation

Deforestation, primarily driven by human activities, is a significant global issue with extensive ecological and socio-economic impacts. Urban forests, which deliver critical ecosystem services, are especially susceptible to these pressures, particularly in areas undergoing rapid urbanization. This section analyzes the causes, effects, and mitigation strategies related to human-induced pressures and deforestation, moving from global to local contexts, and concluding with the case study of the Kazimzumbwi Forest Reserve in Tanzania.

6.2.1 Global anthropogenic pressures and deforestation

Deforestation is a primary factor contributing to biodiversity loss, climate change, and the degradation of ecosystems worldwide. From 2015 to 2020, approximately 10 million hectares of forest were lost each year globally, primarily as a result of human activities (Food and Agriculture Organization [FAO], 2023). Primary factors include:

(i) Agricultural expansion, specifically the clearing of forests for crops and livestock, constitutes approximately 80% of global deforestation (Curtis et al., 2018). The Amazon rainforest is significantly deforested for soybean cultivation and cattle ranching.

(ii) Urbanization results in the swift transformation of forested regions into residential, commercial, and industrial zones. Urban sprawl has led to considerable forest loss in cities such as Jakarta, São Paulo, and Nairobi (Nowak et al., 2021).

(iii) Illegal logging and mining activities lead to forest degradation, especially in areas characterized by weak governance (World Resources Institute [WRI], 2022).

(iv) Infrastructure development, including the construction of roads, dams, and energy facilities, frequently necessitates extensive deforestation, leading to the fragmentation of forest ecosystems and the disruption of wildlife habitats (Global Forest Watch, 2022).

6.2.2 Deforestation and anthropogenic pressures in Africa

Africa experiences significant deforestation challenges attributed to its increasing population and reliance on natural resources. Annually, the continent experiences a loss of roughly 3.9 million hectares of forest, predominantly due to agriculture, energy requirements, and infrastructure expansion (African Forest Forum, 2022).

Significant human-induced pressures encompass:

(i) Charcoal production as in Sub-Saharan Africa, charcoal and firewood serve as the main energy sources for numerous households, resulting in significant deforestation (Chidumayo & Gumbo, 2010).

(ii) Small-scale subsistence farming is a significant contributor to deforestation in nations such as Ghana, Nigeria, and Tanzania (FAO, 2023).

(iii) Urbanization is driving rapid growth, resulting in heightened demand for land and resources, which in turn accelerates forest loss in cities such as Nairobi, Accra, and Dar es Salaam (UNEP, 2022).

(iv) Efforts to address deforestation in Africa encompass reforestation initiatives, agroforestry methods, and the encouragement of alternative energy sources, including solar and biogas, to diminish reliance on wood fuel (Wanjiru & Matsui, 2022).

6.2.3 Sub-Saharan Africa: Forest loss and human activities

Sub-Saharan Africa (SSA) exhibits some of the highest global deforestation rates, primarily due to human activities including agriculture, logging, and infrastructure development. Miombo woodlands, prevalent in much of Sub-Saharan Africa, are especially susceptible to degradation owing to their accessibility and economic significance (Chidumayo & Gumbo, 2010).

Anthropogenic pressures in Sub-Saharan Africa encompass:

(i) Shifting cultivation involves the clearing of forests for temporary agricultural purposes, resulting in forest fragmentation and a decline in biodiversity.

(ii) Fuelwood Harvesting: The reliance on wood as an energy source is significant, with both rural and urban populations depending extensively on forest resources (Global Forest Watch, 2022).

(iii) Urban expansion in cities such as Lagos and Addis Ababa leads to deforestation in peri-urban regions, thereby diminishing ecosystem services essential for urban resilience (Myers et al., 2023).

6.2.4 East Africa: Regional dynamics of deforestation

Deforestation in East Africa represents a critical environmental issue, intensified by factors such as population growth, urbanization, and climate change. Primary factors include:

(i) Agriculture and livestock practices in countries such as Kenya, Uganda, and Tanzania involve the clearing of forests for crop cultivation and grazing, leading to a decline in forest cover and a reduction in ecosystem functionality (Maathai, 2021).

(ii) Infrastructure development, including roads, pipelines, and dams, significantly contributes to deforestation and forest fragmentation, especially in forested areas such as the Eastern Arc Mountains (Nyika, 2020).

(iii)Charcoal production is a significant contributor to forest degradation in East Africa, as urban populations heavily depend on this energy source (UNEP, 2022).

Mitigation of deforestation involves community-based conservation programs, forest restoration initiatives, and the incorporation of sustainable forestry practices into national policies.

6.2.5 Anthropogenic pressures and deforestation in Tanzania

Tanzania undergoes considerable deforestation, with an annual loss of roughly 372,000 hectares of forest. Human-induced pressures, including agriculture, urbanization, and energy demands, are key drivers (Ministry of Natural Resources and Tourism [MNRT], 2022).

(i) Smallholder farming is a primary driver of deforestation, with forests being cleared for the cultivation of crops such as maize and cassava.

(ii) Charcoal production: More than 90% of households in Tanzania depend on charcoal and firewood for cooking, leading to significant deforestation (Mgaya & Samwel, 2021).

(iii) Urban expansion: Cities such as Dar es Salaam and Dodoma are encroaching upon peri-urban forests, endangering vital ecosystems including the Pugu and Kazimzumbwi Forest Reserves (Nyika, 2020).

Conservation initiatives in Tanzania encompass Participatory Forest Management (PFM), reforestation efforts, and the encouragement of alternative energy sources to diminish dependence on forest resources.

6.2.6 Kazimzumbwi Forest Reserve: A case study

The Kazimzumbwi Forest Reserve, located near Dar es Salaam, epitomizes the challenges and opportunities of managing urban forests amid anthropogenic pressures.

S/n	Anthropogenic drivers	Explanation
1.	Urban encroachment	Kazimzumbwi is increasingly under threat from urban expansion, as nearby settlements and infrastructure projects encroach on the forest's boundaries.
2.	Illegal logging	The demand for timber and charcoal has led to unsustainable harvesting practices, degrading the forest and reducing its ecological integrity (Mwakalila et al., 2023).
3.	Agricultural activities	Small-scale farming within and around Kazimzumbwi contributes to deforestation and biodiversity loss
S/n	Impacts	Explanation
1.	Biodiversity loss	The degradation of Kazimzumbwi threatens its unique flora and fauna, reducing the forest's role as a biodiversity hotspot.
2.	Reduced ecosystem services	Deforestation compromises the forest's ability to sequester carbon, regulate temperatures, and provide water catchment services (Mgaya & Samwel, 2021).
S/n	Conservation strategies	Explanation
1.	Community engagement	Local communities are involved in reforestation initiatives and conservation education programs.
2.	Policy enforcement	Strengthened enforcement of forest protection laws aims to curb illegal activities.
3.	Integration into urban planning	Incorporating Kazimzumbwi into Dar es Salaam's urban development plans can enhance its conservation while supporting sustainable urban growth (MNRT, 2022).

CHAPTER SEVEN

7. POLICY FRAMEWORKS AND URBAN PLANNING

7.1 Current Policies and Gaps

Effective forest management policies are crucial for sustaining urban forests and their capacity to provide ecological and socio-economic benefits. Although global, regional, and national frameworks for forest preservation exist, substantial policy gaps impede the achievement of these goals. This section examines existing policies and their limitations, transitioning from global initiatives to the particular instance of the Kazimzumbwi Forest Reserve in Tanzania.

7.1.1 Global forest policies and gaps

Various international policy frameworks focus on forest conservation and sustainable management, such as the United Nations Framework Convention on Climate Change (UNFCCC), the Convention on Biological Diversity (CBD), and the United Nations Forum on Forests (UNFF). These frameworks highlight the necessity of incorporating forest conservation within the wider context of climate, biodiversity, and sustainable development objectives.

The UNFCCC's REDD+ initiative (Reducing Emissions from Deforestation and Forest Degradation) promotes forest conservation in developing countries to serve as carbon sinks. REDD+ has provided funding and technical support; however, its implementation encounters challenges, including insufficient capacity-building and ambiguous land tenure rights in numerous countries (Angelsen et al., 2018).

Sustainable Development Goals (SDGs): SDG 15, titled "Life on Land," underscores the importance of sustainable forest management. Global deforestation rates demonstrate that progress toward this goal remains inadequate (United Nations, 2023).

Despite these frameworks, there are deficiencies in enforcement, financing, and the consideration of socio-economic factors contributing to deforestation. The emphasis on forest carbon sequestration globally frequently neglects urban forests, which play a vital role in addressing urban issues like heat islands and air pollution (FAO, 2023).

7.1.2 Forest policies and gaps in Africa

Africa's forest policies are influenced by the African Union's Agenda 2063 and regional efforts like the African Forest Landscape Restoration Initiative (AFR100). The frameworks seek to rehabilitate degraded landscapes, improve biodiversity, and bolster livelihoods.

Agenda 2063 highlights the importance of sustainable resource management by the African Union; however, it does not include specific components related to urban forestry, which restricts its relevance in rapidly urbanizing regions (African Union, 2022).

AFR100 seeks to rehabilitate 100 million hectares of degraded land by the year 2030. Urban forests are often overlooked in comparison to rural landscapes (FAO, 2023).

In Africa, policy gaps are characterized by weak enforcement mechanisms, inadequate funding, and a lack of sufficient integration of urban forestry into national and regional policies. The significant dependence on external funding raises concerns regarding the sustainability of long-term conservation initiatives (UNEP, 2022).

7.1.3 Sub-Saharan Africa: Policy frameworks and challenges

Sub-Saharan Africa (SSA) contains some of the most biodiverse forests globally; however, these ecosystems face escalating threats from deforestation, urbanization, and climate change. Significant policy frameworks in Sub-Saharan Africa encompass:

National Forest Programs (NFPs) have been adopted by numerous SSA countries, emphasizing the importance of forest conservation and sustainable utilization. These programs frequently lack strategies tailored to urban environments, resulting in increased vulnerability of urban forests to degradation (Global Forest Watch, 2022).

Participatory Forest Management (PFM) initiatives in Sub-Saharan Africa promote community engagement in forest governance. Their effectiveness in rural areas is noted; however, application to urban forests is constrained by competing land-use priorities and insufficient institutional support (Chidumayo & Gumbo, 2010).

Policy deficiencies in Sub-Saharan Africa encompass the omission of urban forests from national strategies, insufficient legal frameworks for the protection of peri-urban forests, and a restricted emphasis on climate resilience within forest policies.

7.1.4 Forest policies and gaps in East Africa

East African nations have advanced in forest conservation via national policies, regional collaborations, and international partnerships. Prominent frameworks consist of:

(i) The Forest Conservation and Management Act of 2016 in Kenya underscores the importance of community involvement and the sustainable utilization of forest resources. Urban forests, such as Karura Forest, are subject to persistent threats from urban expansion and illegal activities (Wanjiru & Matsui, 2022).

(ii) Uganda's National Forest Policy (2001) emphasizes forest restoration and biodiversity conservation; however, it does not provide comprehensive strategies for urban forests (Nyika, 2020).

In East Africa, deficiencies encompass insufficient enforcement of forest regulations, constrained financial resources for urban forest management, and inadequate incorporation of urban forests into climate action strategies. To address these gaps, it is essential to strengthen regional collaboration and enhance stakeholder engagement (UNEP, 2022).

7.1.5 Forest policies and gaps in Tanzania

Tanzania has implemented various policies and frameworks to direct forest conservation and management, such as the National Forest Policy (1998, revised 2022), the Forest Act (2002), and Participatory Forest Management (PFM) initiatives.

The National Forest Policy (2022) underscores the importance of sustainable forest management, the conservation of biodiversity, and the enhancement of climate resilience. Urban forests are often overlooked, lacking comprehensive guidelines for their protection and incorporation into urban planning (Ministry of Natural Resources and Tourism [MNRT], 2022).

Participatory Forest Management (PFM) in Tanzania has demonstrated success in rural regions; however, urban forests such as Kazimzumbwi and Pugu Forest Reserves encounter difficulties stemming from competing land-use demands and insufficient community engagement (Mgaya & Samwel, 2021).

Significant policy deficiencies in Tanzania encompass:

(i) Urban forestry is not explicitly included in national policies, rendering these areas susceptible to encroachment and degradation.

(ii) Inadequate enforcement of forest regulations results in illegal logging, land encroachment, and deforestation.

(iii)Funding constraints significantly impede the execution of conservation programs, especially in peri-urban regions.

7.1.6 Kazimzumbwi Forest Reserve: Policy gaps and recommendations

The Kazimzumbwi Forest Reserve, situated near Dar es Salaam, illustrates the difficulties associated with the management of urban forests under current policy frameworks. Kazimzumbwi, while acknowledged for its ecological and socio-economic significance, encounters considerable threats stemming from insufficient policy focus and enforcement.

Policies currently impacting Kazimzumbwi include land use plans and Forest Act of 2002. Although designated as a forest reserve, Kazimzumbwi is not integrated into the urban planning framework of Dar es Salaam, which heightens its susceptibility to urban encroachment. Forest Act of 2002 as the legislation offers legal safeguards for forest reserves; however, enforcement in peri-urban regions is inadequate, permitting the continuation of illegal logging and agricultural practices.

Deficiencies in policy described as national policies emphasize rural forests, resulting in insufficient protection for peri-urban reserves such as Kazimzumbwi. Also, limited community involvement in conservation efforts diminishes local support and compliance with forest management initiatives. Furthermore, the absence of comprehensive monitoring frameworks impedes the evaluation of forest health and the efficacy of conservation efforts.

Suggestions include incorporating Kazimzumbwi into the urban planning framework of Dar es Salaam can improve its protection and promote sustainable utilization. Moreover, enhance enforcement mechanisms. Enhancing financial support for forest law enforcement and strengthening the capabilities of local authorities can mitigate illegal activities. Likewise, enhance community engagement by promoting participatory conservation approaches and providing alternative livelihoods to strengthen local support for forest preservation.

7.2 Integrating Forests into Sustainable City Planning

Urban forests play a crucial role in the development of sustainable cities, offering significant ecological, economic, and social advantages. Their incorporation into urban planning enhances climate resilience, promotes biodiversity conservation, and improves living conditions in cities. This section examines the incorporation of forests into sustainable urban planning, focusing on a global context with particular emphasis on Africa, Sub-Saharan Africa, East Africa, Tanzania, and the case study of the Kazimzumbwi Forest Reserve.

7.2.1 Global efforts in integrating forests into sustainable cities

The integration of forests into urban planning has become increasingly important as cities confront issues such as urban heat islands, air pollution, and biodiversity loss. Initiatives like the United Nations Sustainable Development Goals (SDG 11: Sustainable Cities and Communities) and the New Urban Agenda highlight the significance of green infrastructure, including forests, in fostering sustainable urban environments (United Nations, 2022).

Fundamental strategies encompasses:

(i) Green infrastructure planning in cities such as Singapore and Copenhagen integrates urban forests to improve climate resilience and overall livability. Singapore's "City in a Garden" strategy incorporates extensive tree planting, green roofs, and urban parks to enhance air quality and mitigate urban temperatures (Tan et al., 2021).

(ii) Global policy frameworks, including the European Green Deal, highlight the significance of urban forests in attaining carbon neutrality and enhancing public health (European Commission, 2022).

(iii) Technological integration of Geographic Information Systems (GIS) and remote sensing technologies is increasingly utilized for mapping and managing urban forests, facilitating their effective incorporation into city planning (Nowak et al., 2021).

Challenges such as competing land uses, limited funding, and inadequate policy enforcement continue to impede the global integration of forests into urban planning, despite advancements made in this area.

7.2.2 Integrating forests into sustainable cities in Africa

In Africa, where urbanization is occurring at one of the fastest rates globally, integrating forests into sustainable city planning is both a necessity and a challenge. Urban forests can address pressing issues such as air pollution, water scarcity, and rising temperatures in cities like Nairobi, Accra, and Lagos (African Development Bank [AfDB], 2022).

S/n	Initiatives and challenges	Explanation
1.	Green corridors	Initiatives such as the conservation efforts in Nairobi's Karura Forest illustrate the feasibility of incorporating forests into urban planning frameworks. Green corridors that connect urban forests to city centers offer both recreational spaces and ecological advantages (Wanjiru & Matsui, 2022).
2.	Policy gaps	Although frameworks such as Agenda 2063 advocate for sustainable urbanization, there is a notable lack of emphasis on urban forests. To address these gaps, it is essential to enhance the integration of urban forestry within national and municipal policies (UNEP, 2022).
3.	Community engagement	Community-based forest management programs are increasingly being adopted in African cities, promoting local stewardship of urban forests. Scaling these initiatives necessitates technical support and funding (FAO, 2023).

7.2.3 Sub-Saharan Africa: Forest integration and urban sustainability

Sub-Saharan Africa (SSA) encounters distinct challenges in incorporating forests into urban planning, attributed to rapid urbanization, informal settlements, and limited resources. Urban forests provide a cost-effective means of improving urban sustainability.

Instances of integration include Addis Ababa, Ethiopia has incorporated urban green spaces into its development strategies, utilizing forests to alleviate flooding, enhance air quality, and offer recreational areas (Myers et al., 2023). Moreover, Kigali, Rwanda in their master plan of Kigali includes significant green belts and urban forests aimed at enhancing biodiversity and mitigating urban heat islands, illustrating the capacity for cities in Sub-Saharan Africa to emphasize sustainable urban planning (Global Forest Watch, 2022).

Obstacles aroused includes the demand for housing and infrastructure frequently takes precedence over the designation of land for forests and green spaces. Also, institutional deficiencies, characterized by ineffective governance and inadequate enforcement, hinder urban forest conservation initiatives. Thus, addressing these challenges necessitates collaboration among multiple stakeholders, the development of innovative financing models, and enhanced integration of urban forestry within national urban policies.

7.2.4 Forests and sustainable cities in East Africa

Urban forests are gaining recognition in sustainable city planning among East African cities, including Nairobi, Kampala, and Dar es Salaam. Forests are essential for mitigating the impacts of climate change, enhancing urban biodiversity, and improving public health.

Optimal approaches include Nairobi, Kenya where Karura Forest exemplifies the integration of forest ecosystems within urban planning frameworks. Administered via public-private partnerships, it offers ecological, recreational, and cultural advantages (Wanjiru & Matsui, 2022). Also, Kampala, Uganda where the Green Action Plan of Kampala integrates urban forests into its strategy to improve resilience against flooding and urban heat, highlighting the role of urban forestry in mitigating climate challenges (Nyika, 2020).

Challenges and opportunities encounter East African cities including a lack of technical expertise, insufficient funding, and the encroachment of forested areas. Increased awareness of the benefits of urban forestry is leading to greater investment in conservation and restoration initiatives.

7.2.5 Integrating forests into sustainable cities in Tanzania

Tanzania has advanced in incorporating forests into sustainable urban planning, especially in cities such as Dar es Salaam, Arusha, and Dodoma. Urban forests, exemplified by the Pugu and Kazimzumbwi Reserves, deliver essential ecosystem services such as carbon sequestration, temperature regulation, and biodiversity conservation (Ministry of Natural Resources and Tourism [MNRT], 2022).

Frameworks for policy include The National Forest Policy (2022) highlights the importance of sustainable forest management; however, it does not provide clear strategies for the incorporation of urban forests within urban planning frameworks. While, current urban planning legislation acknowledges green spaces; however, it frequently neglects urban forests, rendering them susceptible to encroachment and deterioration (Nyika, 2020).

Obstacles encountered include urban encroachment resulting from rapid urbanization in cities such as Dar es Salaam has caused considerable deforestation in peri-urban areas, notably the Kazimzumbwi Forest Reserve. Furthermore, funding gaps significantly restrict the execution of urban forestry initiatives due to insufficient financial resources.

Prospects to tackle the obstacles include participatory approaches that involve local communities can improve urban forest conservation initiatives. Also, enhancing the incorporation of forests into urban development strategies can bolster urban sustainability.

7.2.6 Kazimzumbwi Forest Reserve: A case study

The Kazimzumbwi Forest Reserve, situated near Dar es Salaam, exemplifies the challenges and opportunities associated with the integration of forests into sustainable urban planning. Present function in urban sustainability where Kazimzumbwi functions as an essential ecological buffer, reducing urban heat islands, sequestering carbon, and promoting biodiversity. The proximity to Dar es Salaam highlights its significance in tackling urban issues, including air pollution and water scarcity (Mwakalila et al., 2023).

Obstacles faced include urban expansion and infrastructure development pose significant threats to the integrity of forests. Kazimzumbwi exhibits weak integration within Dar es Salaam's urban planning framework, which constrains its protection and sustainable utilization. Possibilities to solve obstacles include integrating Kazimzumbwi into Dar es Salaam's master plan can improve urban sustainability. Also, Public-Private Partnerships as utilizing partnerships can effectively mobilize resources for conservation and restoration initiatives. Furthermore, the development of eco-tourism initiatives has the potential to generate revenue and enhance public awareness regarding the value of forests.

CHAPTER EIGHT

8. INNOVATIONS IN FOREST MANAGEMENT

8.1 Technology for Monitoring (e.g., Remote Sensing)

Technological advancements have transformed forest monitoring, allowing for accurate tracking of ecological and socio-economic changes in urban forests. Remote sensing, Geographic Information Systems (GIS), and various digital technologies are essential instruments for efficient forest management. These technologies facilitate the monitoring of forest health, assessment of land use and land cover changes, and evaluation of the impacts of human activities on ecosystems. This section examines the role of technology in forest monitoring, focusing on a global perspective and narrowing down to Africa, Sub-Saharan Africa, East Africa, Tanzania, and the Kazimzumbwi Forest Reserve.

8.1.1 Global advances in remote sensing for forest monitoring

Remote sensing technologies are essential for the global monitoring of forests. Satellite imagery, aerial photography, and drone-based sensors offer extensive, high-resolution data that facilitate the evaluation of forest cover, biomass, and carbon stocks.

Utilizations include analysis of land use and land cover change as remote sensing facilitates the detection of deforestation, urban expansion, and forest degradation. The Landsat program by NASA provides multi-decadal data for the global monitoring of forest changes (Hansen et al., 2013). Also, carbon stock estimation as LiDAR (Light Detection and Ranging) technology effectively estimates forest biomass and carbon storage, which are essential for climate change mitigation strategies (Asner et al., 2020). Lastly, forest health monitoring utilizes spectral data from satellites like Sentinel-2 and MODIS to identify stress in vegetation resulting from drought, pests, or pollution, facilitating timely interventions (Zhang et al., 2021).

Obstacles encompasses global forest monitoring technologies, while useful, encounter challenges including data accessibility, the expense of high-resolution imagery, and restricted capacity for ground-truthing in developing areas.

8.1.2 Remote sensing applications in Africa

In Africa, deforestation and land degradation are significant issues, and remote sensing technologies are essential for monitoring forests. Satellite imagery offers essential insights into deforestation trends, especially in areas such as the Congo Basin and the Sahel.

Principal initiatives encompass The African Regional Data Cube initiative employs satellite data to track land use changes, such as deforestation and urbanization, in various African nations (Global Partnership for Sustainable Development Data, 2021). Also, REDD+ Projects as numerous African nations employ remote sensing to fulfill REDD+ (Reducing Emissions from Deforestation and Forest Degradation) reporting obligations, especially in the context of carbon accounting (Angelsen et al., 2018).

Obstacles aroused as despite the value of remote sensing, numerous African nations face limitations in technical capacity and financial resources, hindering their ability to fully leverage these technologies. Educating local stakeholders and improving data accessibility are essential for addressing these challenges.

8.1.3 Sub-Saharan Africa: Leveraging technology for forest monitoring

Sub-Saharan Africa (SSA) faces unique challenges in forest management due to rapid urbanization, population growth, and weak governance. However, remote sensing technologies are increasingly employed to monitor forests and guide conservation strategies.

Applications encompasses early warning systems as remote sensing supports the detection of illegal logging and forest fires in SSA. For instance, tools like Global Forest Watch use satellite imagery to issue alerts for deforestation hotspots (Hansen et al., 2013). Moreover, community monitoring as combining remote sensing with community-based forest management empowers local populations to monitor and manage forests sustainably (Chidumayo & Gumbo, 2010).

Opportunities available include scaling up the use of affordable tools like drones and open-source platforms can enhance forest monitoring across SSA. Integrating remote sensing data into national forest policies can also improve governance and enforcement.

8.1.4 Remote sensing in East Africa

East Africa leads in the application of remote sensing technologies for forest monitoring, especially in Kenya, Uganda, and Tanzania. Urban forests in this region are significantly threatened by deforestation and land-use changes, necessitating technological interventions.

Optimal approaches include satellite data is extensively utilized to map and monitor the mangrove forests of East Africa, which play a crucial role in carbon sequestration and coastal protection (Maathai, 2021). Also, urban forest assessment as remote sensing and GIS technologies are employed in Nairobi to evaluate changes in tree cover and their effects on urban ecosystems (Wanjiru & Matsui, 2022).

Obstacles and advancements in East Africa faces challenges such as insufficient funding for high-resolution imagery and a lack of technical expertise. Initiatives such as the Regional Centre for Mapping of Resources for Development (RCMRD) are addressing these gaps through the provision of training and resources for remote sensing applications.

8.1.5 Remote sensing for forest monitoring in Tanzania

Remote sensing technologies are being increasingly utilized in Tanzania to monitor forests, encompassing urban and peri-urban reserves. These technologies are essential for mitigating deforestation, land degradation, and urban encroachment.

Utilization in forest surveillance as the Tanzanian forest service employs satellite imagery to track alterations in forest cover and fulfill national reporting obligations (Ministry of Natural Resources and Tourism [MNRT], 2022). Moreover, urban forest assessment: remote sensing facilitates the mapping of urban forests in cities such as Dar es Salaam, enabling the identification of areas vulnerable to deforestation and informing conservation strategies (Nyika, 2020).

Obstacles that Tanzania encounters including restricted access to high-resolution data and inadequate technical capacity for data analysis. Enhancing collaborations with international organizations may effectively address these deficiencies.

8.1.6 Kazimzumbwi Forest Reserve: A case study in remote sensing

The Kazimzumbwi Forest Reserve near Dar es Salaam provides a compelling example of the role of remote sensing in urban forest monitoring. Applications include deforestation detection as satellite imagery reveals patterns of deforestation and encroachment in Kazimzumbwi, helping policymakers identify critical areas for intervention (Mwakalila et al., 2023). Also, carbon stock assessment as LiDAR and multispectral imaging are used to estimate the reserve's carbon sequestration potential, supporting climate action goals (Mgaya & Samwel, 2021). Furthermore, biodiversity mapping as remote sensing helps map habitats within Kazimzumbwi, identifying areas critical for biodiversity conservation.

Opportunities for enhancement include community integration as combining remote sensing data with community-based monitoring can enhance local engagement and conservation outcomes. Besides, data accessibility as increasing access to high-resolution data through partnerships with global initiatives like Global Forest Watch can improve the reserve's management. Moreover, policy Integration: Incorporating remote sensing insights into urban planning frameworks can strengthen the reserve's protection against urban encroachment (MNRT, 2022).

8.2 Community-Driven Initiatives

Community-driven initiatives play a crucial role in sustainable forest management, especially in urban and peri-urban contexts. These initiatives utilize local knowledge, encourage collective responsibility, and create opportunities for socio-economic benefits while enhancing ecological integrity. Community-driven initiatives, such as the Kazimzumbwi Forest Reserve in Tanzania, play a crucial role in the preservation and sustainability of urban forests, reflecting both global efforts and localized actions.

8.2.1 Global community-driven forest initiatives

Community participation in forest management is widely acknowledged as essential for sustainable conservation efforts worldwide. Community-based forest management (CBFM) programs prioritize local participation in decision-making, monitoring, and benefit-sharing, thereby enabling communities to safeguard their forest resources.

Principal illustrations include the Green Belt Movement in Kenya, established by Wangari Maathai, engaged communities in the planting of millions of trees, the restoration of degraded landscapes, and the promotion of environmental and social justice. The success underscores the transformative potential of community-led initiatives (Maathai, 2021). Secondly, Nepal's Community Forestry Program empowers local user groups with forest management rights, enhancing stewardship and livelihood outcomes. Deforestation has been significantly reduced, and forest cover has increased (Chhetri et al., 2022). Lastly, community-based REDD+ projects as in nations like Indonesia and Brazil, local communities are essential in executing REDD+ (Reducing Emissions from Deforestation and Forest Degradation) initiatives, emphasizing sustainable forest management while generating carbon credits (Angelsen et al., 2018).

8.2.2 Community-driven initiatives in Africa

Community-driven initiatives in Africa play a crucial role in forest management, specifically in combating deforestation, land degradation, and climate change. These initiatives frequently correspond with participatory forest management frameworks that highlight community ownership and sustainable resource utilization.

Principal initiatives include the African Forest Landscape Restoration Initiative (AFR100) involves community engagement in the restoration of degraded landscapes throughout Africa, with the objective of restoring 100 million hectares of land by 2030. Tree planting and agroforestry projects driven by community involvement are fundamental to this initiative (African Union, 2022). Also, Participatory Forest Management (PFM) has been adopted by numerous African nations, enabling local communities to collaboratively manage forest resources alongside government agencies. PFM in Ethiopia's Bale Mountains has effectively decreased deforestation and improved livelihoods by providing alternative income sources such as beekeeping (Amente et al., 2023).

Obstacles aroused embrace community-driven initiatives in Africa, while successful in certain aspects, encounter challenges including weak governance, insufficient funding, and disputes regarding resource utilization. Resolving these issues necessitates enhanced policy support and capacity-building initiatives.

8.2.3 Sub-Saharan Africa: Community participation in forest conservation

Sub-Saharan Africa (SSA) encompasses a variety of forest ecosystems, including the Congo Basin and the miombo woodlands. Community-driven initiatives are essential for the sustainable management of these ecosystems, considering the region's reliance on forests for livelihoods and ecosystem services.

Instances of achievement include community forestry in the Congo Basin engages local populations in sustainable logging and conservation efforts, facilitating a balance between economic requirements and forest preservation (Global Forest Watch, 2022). Also, community-based fire management programs in the miombo woodlands of

Zambia and Mozambique have effectively reduced wildfire incidents, thereby preserving biodiversity and enhancing carbon storage (Chidumayo & Gumbo, 2010).

Innovations and opportunities encompasses participatory mapping and community monitoring through mobile technology are enhancing community capacity for effective forest management. Expanding these initiatives can increase their effectiveness throughout Sub-Saharan Africa.

8.2.4 Community-driven initiatives in East Africa

Community-driven initiatives in East Africa are essential for tackling the interconnected issues of deforestation and urbanization. Urban forests in this region, including Karura Forest in Nairobi and Mabira Forest in Uganda, illustrate the significance of local involvement in conservation initiatives.

Case analyses include Karura Forest in Kenya is overseen by the Friends of Karura Forest, a community-based initiative that has successfully converted the area into a vibrant urban park, providing ecological, recreational, and economic advantages. Revenue generated from eco-tourism facilitates continuous conservation efforts (Wanjiru & Matsui, 2022). Also, Mabira Forest, Uganda as community organizations in Uganda engage in reforestation, ecotourism, and the sustainable harvesting of non-timber forest products, thereby offering alternative livelihoods and mitigating deforestation (Nyika, 2020).

Obstacles to achievement include insufficient financial resources, inadequate legal frameworks, and a lack of awareness regarding the significance of urban forests. Overcoming these barriers necessitates integrated strategies that connect conservation efforts with overarching urban development objectives.

8.2.5 Community-driven forest conservation in Tanzania

Tanzania has adopted community-driven strategies for forest conservation via Participatory Forest Management (PFM) and Joint Forest Management (JFM) initiatives. These initiatives enable local communities to actively participate in the management of forest resources and to share in the associated benefits.

Significant programmes include conservation in the Eastern Arc Mountains involves community participation in tree planting, sustainable harvesting, and biodiversity monitoring, which supports both ecological preservation and local livelihoods (Ministry of Natural Resources and Tourism [MNRT], 2022). Also, initiatives aimed at sustainable charcoal production and alternative energy sources have effectively diminished deforestation in both rural and urban settings, underscoring the efficacy of community-driven solutions (Mgaya & Samwel, 2021).

Challenges and opportunities aroused as despite notable successes, urban forests such as Kazimzumbwi and Pugu reserves encounter distinct challenges, including encroachment and insufficient community engagement. Enhancing community involvement and incorporating urban forestry into national policies can effectively mitigate these deficiencies.

8.2.6 Kazimzumbwi Forest Reserve: A case study in community-driven conservation

The Kazimzumbwi Forest Reserve, located near Dar es Salaam, exemplifies the significance of community-driven efforts in the conservation of urban forests. Ongoing initiatives include reforestation involve local communities engaging in tree-planting efforts aimed at restoring degraded regions within the reserve, thereby enhancing forest cover and biodiversity. Also, sustainable livelihoods through beekeeping and ecotourism initiatives offer alternative income sources, thereby decreasing reliance on forest resources and enhancing conservation awareness (Mwakalila et al., 2023). Community-based monitoring involves local volunteers who oversee illegal logging and encroachment, thereby strengthening law enforcement and forest protection efforts.

Potential for expansion involves policy support for the integration of community-driven initiatives into urban planning frameworks can enhance sustainability. Also, capacity building through training and resource provision to local communities enhances the effectiveness of conservation efforts. Moreover, enhancing public awareness regarding the ecological and socio-economic significance of Kazimzumbwi can promote increased community support for its conservation.

PART IV: CASE STUDIES

CHAPTER NINE

9. FOREST CITIES AROUND THE WORLD

9.1 Examples from Developed and Developing Nations

Urban forests serve as essential ecological resources, with cities in both developed and developing countries increasingly acknowledging their significance in tackling environmental, social, and economic issues. This section examines various urban forest initiatives from both developed and developing countries, emphasizing their goals, accomplishments, and insights gained. These examples illustrate the role of urban forests in promoting sustainable urban development worldwide, particularly in Sub-Saharan Africa, East Africa, Tanzania, and the Kazimzumbwi Forest Reserve.

9.1.1 Examples from developed nations

9.1.1.1 Singapore: "City in a Garden" initiative

Singapore exemplifies leadership in urban forestry through its "City in a Garden" initiative, which highlights the incorporation of forests within urban planning frameworks. The city has planted more than 3 million trees, created vertical gardens, and developed urban parks to address urban heat islands, enhance biodiversity, and improve air quality (Tan et al., 2021). Singapore's success illustrates the capacity of urban forests to foster sustainable and habitable urban environments.

Essential characteristics:

   (i) Significant vegetation and urban green spaces.
   (ii) Incorporation of trees and green roofs within infrastructure.
   (iii) Community involvement in tree-planting initiatives.
   (iv) Insights for emerging economies as the initiative highlights the significance of robust governmental commitment, strategic long-term planning, and collaboration between public and private sectors in realizing urban forestry objectives.

9.1.1.2 United States: Urban forests in New York City

New York City (NYC) has made substantial investments in urban forestry via the MillionTreesNYC initiative, which sought to plant 1 million trees throughout the city by 2015. The initiative has enhanced air quality, mitigated urban heat, and created recreational spaces for residents (NYC Parks, 2022).

Essential characteristics:

(i) Collaboration among government entities, non-governmental organizations, and local communities.

(ii) The application of GIS technology for the identification of planting sites and the monitoring of tree health.

(iii) Lessons for developing nations as NYC's approach demonstrates the importance of community involvement and technology in managing urban forests effectively.

9.1.1.3 Germany: Urban green spaces in Berlin

Berlin has integrated urban forests and green spaces into its cityscape as part of its sustainable urban development strategy. The city's Tempelhofer Feld, a repurposed airport, is now an expansive green space that enhances biodiversity and provides recreational opportunities (European Commission, 2022).

Essential characteristics:

(i) Utilization of urban spaces for the implementation of green infrastructure through adaptive reuse.

(ii) Emphasis on biodiversity conservation and public engagement.

(iii) Insights for emerging economies as Berlin's adaptive reuse strategy demonstrates the innovative integration of forests and green spaces within urban planning.

9.1.2 Examples from developing nations

9.1.2.1 Kenya: Karura Forest, Nairobi

Karura Forest in Nairobi exemplifies effective urban forest conservation in a developing nation. The Friends of Karura Forest manage this initiative, which has converted the forest into a vibrant urban park that provides ecological, social, and economic advantages (Wanjiru & Matsui, 2022).

Essential characteristics:

(i) Generating revenue via eco-tourism.
(ii) Participation of the community in the management of forest resources.
(iii) Reforestation as a method for restoring degraded areas.
(iv) Insights for other developing countries as Karura Forest exemplifies the effectiveness of community-based strategies in the conservation of urban forests while also providing economic advantages.

9.1.2.2 India: Implementation of Miyawaki Forests in urban environments

India has implemented the Miyawaki method of afforestation in urban regions to establish dense and rapidly growing forests. Mumbai and Bengaluru have adopted this strategy to increase urban green cover, enhance biodiversity, and address pollution (Bharucha & Mandlekar, 2020).

Essential characteristics:

(i) Utilization of indigenous species for the establishment of self-sustaining forest ecosystems.
(ii) Collaboration between municipal authorities and local communities is essential.
(iii) Insights for other developing nations as the Miyawaki method demonstrates the viability of affordable, scalable approaches for urban afforestation in densely populated regions.

9.1.2.3 Tanzania: Participatory forest management in Dar es Salaam

Participatory Forest Management (PFM) has been implemented in Tanzania to conserve urban forests, including the Pugu and Kazimzumbwi reserves. The initiatives engage local communities in reforestation and sustainable resource management, targeting urban deforestation and biodiversity decline (Ministry of Natural Resources and Tourism [MNRT], 2022).

Essential characteristics:

(i) Reforestation initiatives led by community involvement.
(ii) Programs aimed at enhancing capacity for sustainable forest management.
(iii) Insights for other developing nations as Tanzania's Public Financial Management approach emphasizes the significance of incorporating local knowledge and engaging community participation in the conservation of urban forests.

9.1.3 Regional examples from Sub-Saharan Africa

Sub-Saharan Africa encounters distinct challenges in urban forestry, attributed to swift urbanization and constrained resources. Successful examples illustrate the capacity of urban forests to tackle these challenges.

9.1.3.1 Ethiopia: Addis Ababa Green Legacy Initiative

The Green Legacy Initiative in Addis Ababa has facilitated the planting of billions of trees throughout Ethiopia, encompassing urban regions. This initiative tackles deforestation, air pollution, and urban heat, while involving citizens in conservation efforts (African Union, 2022).

Essential characteristics include extensive tree planting initiatives; and incorporating urban forests into municipal planning.

9.1.3.2 Rwanda: Kigali Green Belt Initiative

The Green Belt project in Kigali incorporates forests into urban planning to enhance biodiversity, improve air quality, and reduce flooding risks. This initiative illustrates the integration of urban forestry within comprehensive climate resilience strategies (Global Forest Watch, 2022).

9.1.4 Examples from East Africa

Cities in East Africa, including Nairobi, Kampala, and Dar es Salaam, are implementing urban forestry initiatives to tackle issues such as deforestation, loss of biodiversity, and urban heat. Kampala, Uganda has implemented green action plans that incorporate urban forests as essential elements for climate adaptation and biodiversity conservation (Nyika, 2020). While, Dar es Salaam, Tanzania as the conservation initiatives for peri-urban forests such as Kazimzumbwi in Dar es Salaam highlight the necessity of incorporating forests into urban planning frameworks (Mwakalila et al., 2023).

9.1.5 Case Study: Kazimzumbwi Forest Reserve, Tanzania

The Kazimzumbwi Forest Reserve, situated near Dar es Salaam, functions as an essential ecological buffer and biodiversity hotspot. It, however, confronts considerable challenges from urban encroachment, illegal logging, and resource exploitation. Community-Driven Initiatives Local communities significantly contribute to the conservation of Kazimzumbwi through reforestation, ecotourism, and sustainable resource utilization (Mgaya & Samwel, 2021). Integration into Urban Planning Initiatives to incorporate Kazimzumbwi into Dar es Salaam's urban planning seek to bolster its conservation while facilitating sustainable urban development (MNRT, 2022).

9.2 Comparative Analysis

A comparative analysis of urban forest initiatives offers insights into the variety of approaches and their results, facilitating the identification of best practices, challenges, and areas for enhancement. This section analyzes management and conservation efforts of urban forests at global, African, and local levels, with a focus on the Kazimzumbwi Forest Reserve in Tanzania, highlighting similarities, differences, and lessons learned.

9.2.1 Global urban forest initiatives: Successes and challenges

Urban forest initiatives worldwide exhibit significant variation in scale, focus, and outcomes, influenced by socio-economic contexts, governance structures, and environmental conditions.

Advanced economies as urban forest programs in developed countries frequently prioritize sustainability, community involvement, and climate resilience. Singapore's "City in a Garden" concept incorporates urban forests into city planning to address urban heat, enhance air quality, and promote biodiversity. The success of the initiative can be attributed to long-term planning, effective governance, and public-private partnerships (Tan et al., 2021). Also, the MillionTreesNYC initiative in New York emphasizes tree planting and community engagement to enhance urban ecosystems. The success illustrates the significance of utilizing technology for monitoring and facilitating collaboration among stakeholders (NYC Parks, 2022).

Emerging economies as urban forest initiatives in developing countries are frequently motivated by the necessity to combat deforestation, urban encroachment, and socio-economic issues. Kenya's Karura Forest exemplifies the success of community-based management in transforming an urban area into a flourishing park. This case illustrates the role of local stewardship in conserving urban forests and generating economic advantages through eco-tourism (Wanjiru & Matsui, 2022). Also, the implementation of the Miyawaki method in urban areas such as Mumbai and Bengaluru demonstrates its potential to mitigate air pollution, enhance biodiversity, and expand green spaces at a low cost (Bharucha & Mandlekar, 2020).

Comparative analysis encompasses community involvement as both developed and developing nations emphasize the significance of community engagement. Nonetheless, the extent and organization of participation vary, as developed countries frequently utilize advanced technology and formal collaborations, whereas developing countries depend on grassroots mobilization. Developed countries generally possess greater financial and technical resources for the implementation of large-scale urban forestry programs. Developing nations encounter limitations yet offset these challenges through innovative and cost-effective strategies.

9.2.2 Urban forestry in Africa vs. global efforts

African nations encounter distinct challenges in urban forest conservation, characterized by rapid urbanization, inadequate governance, and constrained resources. Successful initiatives exemplify resilience and innovation.

Essential comparisons include governance and policy as global initiatives typically gain from robust policy frameworks; however, African nations often face challenges related to enforcement and institutional capacity. Singapore's integrated urban planning exemplifies a contrast to the fragmented governance observed in African cities (FAO, 2023). Also, socio-economic integration as initiatives in Africa, exemplified by Tanzania's Participatory Forest Management (PFM), underscore the simultaneous emphasis on conservation and enhancement of livelihoods. This model is less prevalent in developed countries, where urban forests are primarily utilized for recreational and ecological purposes (MNRT, 2022). Furthermore, advanced technologies such as LiDAR and GIS are increasingly utilized in developed countries, facilitating accurate forest monitoring. African nations are increasingly employing satellite imagery and mobile-based community monitoring to tackle resource limitations (Global Forest Watch, 2022).

9.2.3 Sub-Saharan Africa: Regional trends and variations

Sub-Saharan Africa (SSA) exhibits regional diversity in urban forestry practices, influenced by environmental, economic, and cultural factors. Eastern vs. Western Africa varies as Eastern Africa countries like Kenya and Rwanda focus on integrating urban forests into climate resilience strategies. For instance, Kigali's Green Belt project emphasizes biodiversity conservation and flood mitigation (Nyika, 2020). Western

Africa focus on urban forestry in countries like Nigeria and Ghana often addresses socio-economic needs, with initiatives targeting deforestation and promoting sustainable livelihoods (Chidumayo & Gumbo, 2010).

Lessons learned encompasses policy adaptation as SSA's diversity requires tailored policies that account for regional differences in urbanization patterns and resource availability. Also, community empowerment as programmes emphasizing local participation, such as Ethiopia's Green Legacy Initiative, demonstrate the importance of aligning conservation with community needs (African Union, 2022).

9.2.4 East Africa vs. Other Sub-Saharan regions

East Africa exhibits unique methodologies in urban forestry, emphasizing community-led conservation and the incorporation of forests within urban planning frameworks. Strengths in East Africa community engagement as initiatives such as Kenya's Karura Forest and Tanzania's Participatory Forest Management (PFM) highlight the region's commitment to local involvement and collaborative management. Also, climate resilience as urban forests are increasingly acknowledged for their capacity to mitigate climate risks, as evidenced by Uganda's Green Action Plan and Dar es Salaam's initiatives to safeguard peri-urban reserves (Nyika, 2020). Challenges in the region encroachment and deforestation as the rapid pace of urbanization poses significant threats to urban forests, underscoring the need for enhanced enforcement and better integration into urban planning. Limited Technology Use: Despite some advancements, the uptake of sophisticated monitoring technologies remains uneven.

9.2.5 Tanzania vs. global and regional trends

Tanzania's strategy for urban forestry incorporates global best practices while addressing specific local challenges. The conservation initiatives in the Pugu and Kazimzumbwi reserves highlight the importance of participatory governance and the incorporation of socio-economic benefits into conservation efforts.

Analysis of comparisons encompasses policy integration: asTanzania's National Forest Policy (2022) promotes sustainable forest management; however, urban forests are comparatively underrepresented relative to global benchmarks such as Singapore. Enhancing policy integration is essential (MNRT, 2022). Also, Tanzania effectively utilizes community involvement through initiatives such as Participatory Forest Management (PFM), which corresponds with regional trends in Sub-Saharan Africa.

Areas for enhancement include integrating technology through the expanded application of GIS and remote sensing can improve forest monitoring and management. Also, integrating urban forests into city master plans can alleviate the effects of urbanization and climate change.

9.2.6 Case Study: Kazimzumbwi Forest Reserve in comparative context

The Kazimzumbwi Forest Reserve, situated near Dar es Salaam, exemplifies the urban forestry challenges and opportunities present in Tanzania. Strengths community engagement as local communities participate in reforestation and conservation efforts, reflecting successful models such as Karura Forest in Kenya. Socio-economic benefits as initiatives like sustainable charcoal production and eco-tourism provide income while alleviating pressure on forest resources (Mgaya & Samwel, 2021). Weaknesses includes policy gaps as in contrast to Singapore's integrated planning model, Kazimzumbwi faces weak policy enforcement and limited integration into urban development frameworks. Also, resource constraints as funding and technical capacity pose significant obstacles to effective conservation. Lessons learned in Kazimzumbwi highlights the necessity for multi-stakeholder collaboration, integrating global best practices with local innovations to tackle urban forestry challenges in developing countries.

CHAPTER TEN

10. LESSONS AND RECOMMENDATIONS

10.1 Synthesis of Findings

Urban forests are vital elements of sustainable cities, playing a significant role in ecological equilibrium and socio-economic advancement. This synthesis incorporates findings from global, African, and local perspectives, emphasizing the Kazimzumbwi Forest Reserve in Tanzania. This analysis emphasizes critical insights, shared characteristics, challenges, and opportunities in utilizing the ecological and socio-economic values of urban forests.

10.1.1 Global perspectives: Ecological and socio-economic integrities

Urban forests play a crucial role in mitigating environmental issues, including climate change, air pollution, and biodiversity loss on a global scale. Initiatives such as Singapore's City in a Garden and New York's MillionTreesNYC demonstrate the significant impact of incorporating forests into urban environments (Tan et al., 2021; NYC Parks, 2022).

Principal conclusions indicates urban forests provide ecological benefits by mitigating urban heat islands, sequestering carbon, enhancing air quality, and supporting urban biodiversity. Also, forests contribute to socio-economic development by offering recreational areas, improving mental health, and creating economic opportunities via eco-tourism and sustainable resource management. Furthermore, integration of policy and technology as advanced technologies, including GIS and remote sensing, improve the monitoring and management of urban forests, while thorough urban planning incorporates green infrastructure.

Obstacles include conflicting land-use priorities and urban expansion pose significant risks to forest conservation. Also, financial and institutional limitations hinder the effectiveness of long-term forest management initiatives.

10.1.2 Africa: Leveraging forests amid urbanization

Africa's swift urbanization poses distinct challenges for the management of urban forests. Initiatives such as Kenya's Karura Forest and Ethiopia's Green Legacy Initiative illustrate the continent's capacity to combat deforestation and urban ecosystem degradation (Maathai, 2021; African Union, 2022).

Principal conclusions indicate African initiatives prioritize community engagement, integrating traditional practices with contemporary conservation techniques. Besides, urban forests enhance ecological stability while offering alternative livelihoods, thereby decreasing reliance on unsustainable practices.

Obstacles encountered include inadequate governance and enforcement mechanisms impede the efficacy of forest protection efforts. Moreover, resource constraints hinder the expansion of urban forestry programmes.

10.1.3 Sub-Saharan Africa: Regional trends

In Sub-Saharan Africa (SSA), urban forests are critical for mitigating climate risks, enhancing biodiversity, and supporting livelihoods. Efforts like community-based forest management in the Congo Basin highlight the importance of integrating ecological and socio-economic priorities (Chidumayo & Gumbo, 2010). Key findings include ecosystem services as urban forests in SSA regulate water cycles, improve soil stability, and act as carbon sinks. Also, socio-economic impact as initiatives such as sustainable charcoal production and agroforestry provide economic benefits to local communities. Challenges encountered include encroachment and deforestation driven by urban expansion; and limited access to technology for monitoring and managing urban forests.

10.1.4 East Africa: Promising approaches

Kenya, Rwanda, and Tanzania exemplify innovative approaches to urban forest conservation in East Africa. Examples include Karura Forest in Nairobi and the Green Belt Project in Kigali, which incorporate forests into urban planning to enhance climate resilience and conserve biodiversity (Nyika, 2020; Wanjiru & Matsui, 2022).

Principal conclusions focused on integrated planning as urban forests are being increasingly integrated into the master plans of East African cities to promote urban sustainability. Also, participatory forest management enhances local stewardship and aligns conservation objectives with community interests.

Challenges encountered inconsistent enforcement of policies and limitations in funding. Also, insufficient public awareness concerning the advantages of urban forests.

10.1.5 Tanzania: Conservation and urban integration

Urban forests in Tanzania, including the Pugu and Kazimzumbwi reserves, are essential for ecological preservation and socio-economic development. Participatory Forest Management (PFM) has proven to be an effective strategy; however, challenges remain due to urban pressures (Mgaya & Samwel, 2021).

Principal conclusions cover the National Forest Policy of Tanzania (2022) establishes a basis for sustainable forest management; however, it requires enhanced integration of urban forestry. Also, local communities participate in conservation efforts via reforestation, eco-tourism, and sustainable resource utilization.

Obstacles includes urban encroachment and illegal logging pose significant threats to forest ecosystems. Also, financial and technological constraints hinder conservation initiatives.

10.1.6 Kazimzumbwi Forest Reserve: A microcosm of challenges and opportunities

The Kazimzumbwi Forest Reserve, located near Dar es Salaam, illustrates both the opportunities and difficulties associated with urban forest conservation in rapidly urbanizing environments. The peri-urban forest offers critical ecosystem services but is subjected to considerable anthropogenic pressures (Mwakalila et al., 2023).

Principal conclusions focus on Kazimzumbwi functions as a carbon sink, enhances biodiversity, and alleviates urban heat islands. Also, community role as local initiatives such as tree planting and community monitoring play a crucial role in conservation efforts. Challenges encountered include encroachment as the ecological integrity of the forest is threatened by rapid urbanization and agricultural expansion. Also, deficiencies in policy as inadequate enforcement of conservation policies restricts the protection of the forest.

10.1.7 Common themes across scales

Various themes regarding the integrity of urban forests emerge across global and local contexts. Urban forests universally enhance biodiversity, sequester carbon, and regulate microclimates, contributing significantly to ecosystem services. Successful initiatives in community engagement emphasize local participation and integrate conservation efforts with socio-economic advantages. Effective policies and technological instruments are crucial for the monitoring, management, and integration of urban forests within urban planning frameworks.

Thus, this synthesis highlights the ecological and socio-economic significance of urban forests. The findings emphasize the significance of strong policies, community engagement, and sustainable practices in the conservation of urban forests, particularly in relation to global initiatives and the specific challenges and opportunities present in Kazimzumbwi Forest Reserve. Addressing challenges such as urban encroachment, inadequate governance, and funding constraints is essential for the continued vitality of urban forests as integral elements of sustainable cities.

10.2 Policy and Community Recommendations for the Future

Urban forests play a vital role in mitigating environmental, social, and economic issues within urban areas. With the rapid acceleration of urbanization worldwide, particularly in developing areas like Africa, it is essential to develop proactive policies and encourage community engagement to maintain the ecological and socio-economic integrity of urban forests. This section presents policy and community recommendations from a global viewpoint, advancing through Africa, Sub-Saharan Africa, East Africa, Tanzania, and focusing on the Kazimzumbwi Forest Reserve case.

10.2.1 Global policy and community recommendations

Urban forest initiatives worldwide are enhanced by strong policies, innovative technologies, and engaged community involvement. Significant challenges persist, including competing land-use priorities, constrained funding, and climate-induced pressures on urban ecosystems.

Recommendations for policy implementation include integration of urban forests into national and city-level development plans is essential for their prioritization in urban planning. Policies such as Singapore's City in a Garden serve as a model for urban areas to integrate development with green infrastructure (Tan et al., 2021). Also, global collaboration and knowledge sharing are essential. Platforms such as the UN's Sustainable Development Goals (SDGs) and the New Urban Agenda ought to facilitate the exchange of best practices in urban forestry (United Nations, 2022). Moreover, governments and international organizations must establish enduring funding sources via public-private partnerships, carbon credits, and eco-tourism.

Recommendations from the community are to involve communities in decision making as participatory methods that engage local stakeholders in the planning and monitoring of urban forests can enhance conservation results. Also, education and awareness initiatives should prioritize informing urban populations about the advantages of forests for biodiversity, climate resilience, and public health.

10.2.2 Policy and community recommendations in Africa

Africa's urban forests encounter challenges including inadequate governance, deforestation, and accelerated urbanization. Community-driven initiatives and participatory management models illustrate the potential for transformative change.

Recommendations for policy implementation are African nations should implement enforceable policies to safeguard urban forests against encroachment and degradation. Kenya's Forest Conservation and Management Act (2016) provides a thorough framework for the protection of forests (Wanjiru & Matsui, 2022). Also, incorporate urban forestry into climate strategies as urban forests must be integrated into Nationally Determined Contributions (NDCs) as outlined in the Paris Agreement, highlighting their significance in climate adaptation and mitigation (African Union, 2022). Moreover, regional collaboration through initiatives such as Agenda 2063 by the African Union can facilitate cross-border partnerships aimed at resource sharing and expertise in urban forest management.

Recommendations from the community include enhance local communities: community-based forest management (CBFM) programmes must offer technical and financial assistance to local organizations, facilitating their engagement in conservation efforts. Also, encourage alternative livelihoods through initiatives such as sustainable charcoal production and eco-tourism, which can diminish reliance on forest resources while aligning economic advantages with conservation objectives (FAO, 2023).

10.2.3 Sub-Saharan Africa: Bridging policy and practice

Sub-Saharan Africa (SSA) has exhibited innovative community-based strategies for urban forestry; however, it continues to face challenges related to policy enforcement and resource constraints.

Recommendations for policy implementation are to incorporate urban forestry into development plans as countries in Sub-Saharan Africa should require the integration of urban forests into city master plans to safeguard green spaces during periods of rapid urbanization. Also, encourage Private Sector Involvement: Implementing tax breaks, subsidies, and incentives for companies that invest in urban forestry can enhance resource mobilization for conservation efforts. Moreover, utilize Technology for Monitoring: Investment in GIS, satellite imagery, and drone technology enhances forest monitoring and management throughout Sub-Saharan Africa (Global Forest Watch, 2022).

Recommendations from the community are to encourage community stewardship: policies must acknowledge local communities as essential stakeholders, granting them the authority to manage and derive benefits from urban forests (Chidumayo & Gumbo, 2010). Also, enhance capacity as training programs for community members in sustainable forest management can improve their effectiveness in contributions.

10.2.4 East Africa: Specific strategies for urban forestry

East Africa is progressively integrating urban forestry into its sustainability and resilience frameworks. Initiatives such as Nairobi's Karura Forest and Kigali's Green Belt exemplify effective practices within the region.

Recommendations for policy development are to establish urban forestry departments as it is essential for governments to create specialized urban forestry departments within city planning authorities to prioritize the integration of green spaces in urban development (Nyika, 2020). Also, policy harmonization involves aligning forestry policies with urban planning and climate policies to ensure cohesive and comprehensive action. Moreover, enhance funding opportunities as regional initiatives, such as the East African community climate change master Plan, ought to prioritize urban forestry in their funding distributions.

Recommendations from the community include promote Public-Private Partnerships (PPPs) as collaborating with local businesses and NGOs can enhance resource mobilization for urban forestry initiatives. Also, enhance public engagement as local governments ought to establish platforms for community members to express their concerns and engage in the decision-making process.

10.2.5 Tanzania: Policy and community actions

Tanzania has established a framework for sustainable forest management via its National Forest Policy (2022) and Participatory Forest Management (PFM) initiatives. Urban forests, such as the Kazimzumbwi Reserve, necessitate targeted interventions.

Recommendations for policy implementation include Tanzania should formulate targeted urban forestry policies that incorporate forests into urban planning frameworks, especially in Dar es Salaam. Also, strengthening enforcement mechanisms to address illegal logging and encroachment is essential for the protection of reserves such as Kazimzumbwi (MNRT, 2022). Besides, integrate urban forests into climate resilience strategies as urban forests must be incorporated into planning to mitigate urban heat islands and flooding risks.

Recommendations from the community include enhancing local engagement as the expansion of Participatory Forest Management (PFM) to peri-urban forests may improve community participation in regions such as Kazimzumbwi (Mgaya & Samwel, 2021). Also, diversification of livelihoods as facilitating alternative income-generating activities, Moreover, practices like sustainable farming, beekeeping, and eco-tourism can diminish reliance on forest resources and strengthen local economic resilience.

10.2.6 Kazimzumbwi Forest Reserve: Targeted recommendations

The Kazimzumbwi Forest Reserve encounters distinct challenges due to urban encroachment, illegal activities, and insufficient incorporation into urban planning. Customized policies and community strategies are crucial for conservation efforts.

Recommendations for policy implementation are to integrate kazimzumbwi into urban plans by incorporating the reserve into Dar es Salaam's city master plan will protect its ecological function while aligning with urban development objectives (Mwakalila et al., 2023). Also, zoning policies should be implemented to designate specific areas for conservation, recreation, and limited sustainable use, thereby balancing ecological and socio-economic needs. Moreover, establish a dedicated fund for the restoration and management of Kazimzumbwi to support reforestation and community-based initiatives.

Recommendations from the community are to promote awareness and responsibility as awareness initiatives can inform communities regarding the ecological and socio-economic significance of Kazimzumbwi, encouraging local stewardship. Also, enhance community monitoring programmes by educating local residents to assess forest health and document illegal activities can strengthen enforcement and promote accountability. Moreover, encourage Collaborative Eco-Tourism: Establishing eco-tourism initiatives in collaboration with local communities can yield financial benefits while enhancing conservation awareness.

Thus, yhe sustainability of urban forests relies on the integration of effective policies and active community participation. Urban forestry initiatives should prioritize ecological integrity, socio-economic benefits, and climate resilience across both global and local scales. Implementing these recommendations will enable cities to sustain urban forests such as the Kazimzumbwi Reserve as essential ecological and social resources.

REFERENCES

Adger, W. N., et al. (2003). Adaptation to climate change in the developing world. Progress in Development Studies, 3(3), 179–195.

Adhikari, B., Di Falco, S., & Lovett, J. C. (2014). Institutions and collective action: Does heterogeneity matter in community-based resource management? World Development, 36(5), 758-782.

African Development Bank (AfDB). (2022). Green growth and sustainable development in Africa. Retrieved from https://www.afdb.org

African Forest Forum. (2022). State of Africa's forests. Retrieved from https://www.afforum.org

African Forest Forum. (2022). State of Africa's forests. Retrieved from https://www.afforum.org

African Union. (2022). Agenda 2063: The Africa we want. Retrieved from https://www.au.int

Allen, C. D., Breshears, D. D., & McDowell, N. G. (2015). On underestimation of global vulnerability to tree mortality and forest die-off from hotter drought in the Anthropocene. Ecosphere, 6(8), 129.

Angelsen, A. (2009). Realizing REDD+: National strategy and policy options. CIFOR.

Angelsen, A., Martius, C., & de Sy, V. (2018). Transforming REDD+: Lessons and new directions. Center for International Forestry Research.

Asner, G. P., Brodrick, P. G., & Anderson, C. B. (2020). Remote sensing for biodiversity and conservation. Science Advances, 6(36), eabc7188.

Bateman, I. J., et al. (2013). Bringing ecosystem services into economic decision-making: Land use in the United Kingdom. Science, 341(6141), 45-50.

Bharucha, E., & Mandlekar, M. (2020). Miyawaki forests: Rapid afforestation in urban areas. Urban Forestry Journal, 58(3), 134–140.

Breiman, L. (2001). Random forests. Machine Learning, 45(1), 5–32.

Bunyangha, J., Mbabazi, J., & Kamanzi, J. (2021). Assessing land use and cover change accuracy: Techniques and challenges. Environmental Monitoring and Assessment, 193(5), 234.

Chander, G., Markham, B. L., & Helder, D. L. (2009). Summary of current radiometric calibration coefficients for Landsat MSS, TM, ETM+, and EO-1 ALI sensors. Remote Sensing of Environment, 113(5), 893–903.

Chapin, F. S., Kofinas, G. P., & Folke, C. (2020). Principles of ecosystem stewardship: Resilience-based natural resource management in a changing world. Springer.

Chapin, F. S., Matson, P. A., & Mooney, H. A. (2000). Principles of Terrestrial Ecosystem Ecology. Springer.

Chave, J., Rejou-Mechain, M., Burquez, A., Chidumayo, E., Colgan, M. S., & Delitti, W. B. (2014). Improved allometric models to estimate the aboveground biomass of tropical trees. Global Change Biology, 20(10), 3177–3190.

Chazdon, R. L., et al. (2020). Restoration in the context of the UN Decade on Ecosystem Restoration. Science Advances, 6(36), eaaz9464.

Chhetri, B. B., Lund, J. F., & Nielsen, T. (2022). Community forest management in Nepal: Lessons for global conservation. Forest Policy and Economics, 122, 102364.

Chidumayo, E. N., & Gumbo, D. J. (2010). The dry forests and woodlands of Africa: Managing for products and services. Earthscan.

Clerici, N., Paracchini, M. L., & Maes, J. (2019). Land-use scenarios for ecosystem services modeling: A methodological framework for integration into policy and planning. Land Use Policy, 83, 160–171.

Congalton, R. G. (1991). A review of assessing the accuracy of classifications of remotely sensed data. Remote Sensing of Environment, 37(1), 35–46.

Congalton, R. G., & Green, K. (2019). Assessing the accuracy of remotely sensed data: Principles and practices (3rd ed.). CRC Press.

Costanza, R., et al. (1997). The value of the world's ecosystem services and natural capital. Nature, 387(6630), 253-260.

Costanza, R., et al. (2014). Changes in the global value of ecosystem services. Global Environmental Change, 26, 152-158.

Curtis, P. G., Slay, C. M., & Harris, N. L. (2018). Classifying drivers of global forest loss. Science, 361(6407), 1108–1111.

de Groot, R., et al. (2012). Global estimates of the value of ecosystems and their services in monetary units. Ecosystem Services, 1(1), 50-61.

Eastman, J. R. (2009). TerrSet Geospatial Monitoring and Modeling System. Clark University Press.

Eastman, J. R. (2016). Transition modeling and scenario prediction in land use dynamics. International Journal of Geospatial Science, 12(4), 220–240.

Engel, S., Pagiola, S., & Wunder, S. (2008). Designing payments for environmental services in theory and practice: An overview of the issues. Ecological Economics, 65(4), 663-674.

European Commission. (2022). The European Green Deal. Retrieved from https://ec.europa.eu

FAO. (2018). The State of the World's Forests 2018. Rome: Food and Agriculture Organization of the United Nations.

Food and Agriculture Organization (FAO). (2022). The state of the world's forests. Retrieved from https://www.fao.org

Food and Agriculture Organization (FAO). (2023). The state of the world's forests. Retrieved from https://www.fao.org

Giri, C. (Ed.). (2012). Remote sensing of land use and land cover: Principles and applications. CRC Press.

Global Forest Watch. (2022). Forest loss in Sub-Saharan Africa. Retrieved from https://www.globalforestwatch.org

Global Partnership for Sustainable Development Data. (2021). African Regional Data Cube. Retrieved from https://www.data4sdgs.org

Gyamfi-Ampadu, E., Wang, C., & Xiao, W. (2020). Land use and land cover change detection in Kumasi Metropolis of Ghana using Landsat satellite imageries. International Journal of Remote Sensing, 41(1), 1–22.

Gyamfi-Ampadu, E., Wang, C., & Xiao, W. (2020). Urban forest dynamics in Ghana: A case study of Kumasi Metropolis. Journal of Urban Ecology, 4(2), 98–113.

Hansen, M. C., Potapov, P. V., Moore, R., Hancher, M., Turubanova, S., & Tyukavina, A. (2021). The mental health benefits of forest exposure. Nature Sustainability, 4(3), 210–217.

Hansen, M. C., Potapov, P. V., Moore, R., Hancher, M., Turubanova, S., & Tyukavina, A. (2021). Global forest change from 2000–2020. Science, 342(6160), 850-853.

IPCC. (2021). Climate Change 2021: The Physical Science Basis. Contribution of Working Group I to the Sixth Assessment Report of the Intergovernmental Panel on Climate Change.

Islami, T., Nugroho, W. T., & Suryadi, S. (2022). Evaluating LULC classifications with kappa index. Remote Sensing Applications, 12(3), 245–257.

Jensen, J. R. (2015). Introductory digital image processing: A remote sensing perspective (4th ed.). Pearson.

Kajembe, G. C., & Malimbwi, R. E. (2021). The evolution of forest management in Tanzania: Lessons from the past. Tanzania Journal of Forestry and Nature Conservation, 14(1), 12–24.

Kalaiyarasi, R., Chandran, S., & Kumar, P. (2023). Post-classification comparison for LULC change detection in forest ecosystems. Journal of Remote Sensing and GIS, 18(1), 102–117.

Kilahama, F. B. (2021). The role of coastal forests in Tanzania's urban ecosystems. African Forest Forum, 18(3), 45–60.

Konijnendijk, C. C. (2008). The forest and the city: The cultural landscape of urban woodland. Springer Science & Business Media.

Konijnendijk, C. C., Nilsson, K., Randrup, T. B., & Schipperijn, J. (2022). Urban forests and trees: A reference book. Springer.

Latella, S., Vecchio, V., & Del Monte, S. (2021). Comparative performance of random forests in LULC classification: A case study. International Journal of Environmental Sciences, 19(2), 87–97.

Lawrence, H. W. (2020). City trees: A historical geography from the Renaissance through the nineteenth century. University of Virginia Press.

Leta, G., Tadele, S., & Dinku, A. (2021). The role of predictive modeling in LULC dynamics. Spatial Sciences Review, 16(4), 56–70.

Lillesand, T., Kiefer, R. W., & Chipman, J. (2015). Remote sensing and image interpretation (7th ed.). Wiley.

Liu, Z., Lin, Y., & Yang, W. (2020). Analyzing land cover change rates in tropical forests. Forestry and Remote Sensing, 14(3), 50–67.

Loughner, C. P., et al. (2012). Roles of urban tree canopy and forests in reducing urban heat islands. Environmental Pollution, 159(8-9), 2129-2138.

Lu, D., Mausel, P., Brondízio, E., & Moran, E. (2004). Change detection techniques. International Journal of Remote Sensing, 25(12), 2365–2407.

Maathai, W. (2021). The green belt movement: Sharing the approach and the experience. Lantern Books.

Mansour, S., Al-Belushi, M., & Al-Awadhi, T. (2023). Seasonal effects on land cover classification accuracy in arid regions using Landsat data. Remote Sensing, 15(3), 600.

Massey, R., Brown, S., & Dent, B. (2023). Monitoring forest cover change using Landsat time series data: A case study from Southeast Asia. Forests, 14(2), 300.

McDonald, R. I., et al. (2020). The importance of urban forests for climate change adaptation and mitigation. Nature Climate Change, 10(5), 334-339.

McPherson, E. G., et al. (2017). Urban forests and carbon storage: The role of public lands. Urban Forestry & Urban Greening, 21, 98-107.

Mgaya, J. E., & Samwel, R. L. (2021). Livelihoods and the ecological role of coastal forests in Tanzania: A case study of Kazimzumbwi Forest Reserve. Journal of Environment and Development, 30(2), 112–126.

Ministry of Natural Resources and Tourism (MNRT). (2022). National Forest Policy 2022. Dar es Salaam, Tanzania: Government Press.

Mishra, A., & Rai, P. (2016). Neural network approaches for LULC classification: A critical review. Remote Sensing Applications, 8(2), 101–112.

Moshi, E., & Kweka, L. (2024). Challenges in sustainable forest management: A Tanzanian perspective. Tanzania Journal of Environmental Studies, 19(3), 97–112.

Mwakalila, A., Samson, E., & Lugeye, B. (2023). Challenges and opportunities in managing Kazimzumbwi Forest Reserve. Eastern African Environmental Review, 29(4), 89–104.

Myers, N., Mittermeier, R. A., Mittermeier, C. G., Fonseca, G. A., & Kent, J. (2023). Biodiversity hotspots for conservation priorities. Nature, 403(6772), 853–858.

Nduati, M., & Kamau, J. (2020). Dynamics of forest cover in Nairobi Arboretum: Implications for urban planning and conservation. African Journal of Ecology, 58(4), 712–725.

Nowak, D. J., & Greenfield, E. J. (2022). Urban forests and their role in mitigating urban heat islands. Urban Forestry & Urban Greening, 58(3), 126–139.

Nowak, D. J., et al. (2020). Carbon storage and sequestration by urban forests in the USA. Environmental Pollution, 178, 229-236.

Nowak, D. J., Hirabayashi, S., Bodine, A., & Greenfield, E. (2023). Tree and forest effects on air quality and human health in urban areas. Environmental Pollution, 256(4), 113–125. https://doi.org/10.1016/j.envpol.2022.113125

NYC Parks. (2022). MillionTreesNYC: Growing the urban forest. Retrieved from https://www.nycgovparks.org

Nyika, J. D. (2020). Urban forests in Tanzania: A review of their socio-economic and ecological contributions. African Journal of Urban Studies, 15(1), 67–81.

Pan, Y., et al. (2011). A large and persistent carbon sink in the world's forests. Science, 333(6045), 988-993.

Pimm, S. L., et al. (2014). The biodiversity of species and their rates of extinction, distribution, and protection. Science, 344(6187), 1246752.

Pinto, J., Silva, M., & Campos, T. (2020). Terrain correction techniques for multitemporal satellite imagery. Remote Sensing Today, 15(2), 203–215.

Puyravaud, J. P. (2003). Standardizing the calculation of the annual rate of deforestation. Forest Ecology and Management, 177(1–3), 593–596.

Qacami, A., Berisha, F., & Krasniqi, B. (2023). Enhancing land change models through machine learning techniques. Remote Sensing and Spatial Modeling, 17(1), 45–60.

Regasa, M. S., & Nones, M. (2021). Accuracy assessment techniques in land use classification. International Journal of GIS Studies, 20(1), 120–132.

Regasa, M. S., & Nones, M. (2021). Modeling the impacts of socio-economic factors on land use transitions. International Journal of GIS Studies, 20(1), 120–132.

Regasa, M. S., & Nones, M. (2022). Transition potential mapping and future LULC projections. Environmental Change Monitoring, 19(2), 130–145.

Rwanga, S. S., & Ndambuki, J. M. (2017). Accuracy assessment of LULC classification using remote sensing and GIS. International Journal of Geosciences, 8(4), 611–622.

Saatchi, S. S., Harris, N. L., Brown, S., Lefsky, M., Mitchard, E. T., & Salas, W. (2021). Benchmark map of forest carbon stocks in tropical regions. Nature Climate Change, 11(6), 563–570.

Seidl, R., Thom, D., & Kautz, M. (2017). Forest disturbances under climate change. Nature Climate Change, 7(6), 395–402.

Seto, K. C., Güneralp, B., & Hutyra, L. R. (2012). Global forecasts of urban expansion to 2030 and direct impacts on biodiversity and carbon pools. Proceedings of the National Academy of Sciences, 109(40), 16083-16088.

Song, C., Woodcock, C. E., Seto, K. C., Lenney, M. P., & Macomber, S. A. (2001). Classification and change detection using Landsat TM data: When and how to correct atmospheric effects? Remote Sensing of Environment, 75(2), 230–244.

Stern, N. (2006). The Economics of Climate Change: The Stern Review. Cambridge University Press.

Tan, P. Y., et al. (2021). Urban greening and climate action: Case studies from Singapore and beyond. Nature Climate Solutions, 12(2), 145–152.

UN. (2015). Transforming our world: The 2030 agenda for sustainable development. United Nations.

UN-Habitat. (2022). Urbanization and sustainable development in Africa. Retrieved from https://www.unhabitat.org

United Nations Environment Programme (UNEP). (2022). Africa's green cities: Strategies for sustainable urban development. Retrieved from https://www.unep.org

United Nations Environment Programme (UNEP). (2022). Global forest outlook. Retrieved from https://www.unep.org

United Nations. (1976). Report of Habitat: United Nations Conference on Human Settlements. Retrieved from https://www.un.org

United Nations. (2022). Sustainable development goals: Goal 11 - Sustainable cities and communities. Retrieved from https://www.un.org

United Nations. (2022). World urbanization prospects. Retrieved from https://www.un.org

Vermote, E., Justice, C., Claverie, M., & Franch, B. (2016). Preliminary analysis of the performance of the Landsat 8/OLI land surface reflectance product. Remote Sensing of Environment, 185, 46–56.

Wale, M., Negash, T., & Demissie, H. (2023). The effectiveness of random forests in LULC analysis for African ecosystems. Journal of Earth Observation, 28(3), 89–99.

Wanjiru, C., & Matsui, K. (2022). Community-based urban forestry initiatives in Nairobi, Kenya. Urban Forestry & Urban Greening, 58(3), 126–139.

World Resources Institute (WRI). (2022). Forests in Africa: Challenges and opportunities. Retrieved from https://www.wri.org

Yirdaw, E., Alemayehu, G., & Mekonnen, D. (2020). Encroachment into Ethiopian protected forests: Drivers and consequences. Ethiopian Journal of Environmental Studies, 13(1), 45–60.

Zhang, H., et al. (2021). Remote sensing applications in monitoring urban forest health. Forest Ecology and Management, 489(1), 118920.

Zhao, C., et al. (2016). Urban ecological sustainability: A case study of the Beijing-Tianjin-Hebei Region. Sustainability, 8(12), 1273.

Zhao, F., & Zhu, Y. (2022). High-resolution satellite imagery for LULC classification: Advances and applications. Journal of Remote Sensing, 40(2), 220–234.